Integrated Pest Management
for Collections

Integrated Pest Management for Collections

Proceedings of 2001: A Pest Odyssey

EDITED BY: Helen Kingsley, David Pinniger, Amber Xavier-Rowe, Peter Winsor

science
museum

ENGLISH HERITAGE

NATIONAL
PRESERVATION
OFFICE

JAMES
JAMES

Published by James & James (Science Publishers) Ltd
35–37 William Road, London, NW1 3ER, UK

A catalogue record for this book is available from the
British Library.

ISBN 1 902916 27 1

Printed in the UK by Hobbs the Printers Ltd

Cover photos courtesy of English Heritage (front cover)
and David Pinniger (back cover)

CONTENTS

Acknowledgements

This conference grew out of a discussion, during a train journey, about the need to advertise more widely the benefits of integrated pest management. Following considerable work by the partner organisations, the idea developed into a plan for an international conference. Special thanks are offered to Alison Walker and Belinda Sanderson from the National Preservation Office for hosting and organising the conference, to Helen Kingsley from the Science Museum and David Pinniger for editing and collating the conference pre-prints, to Robert Child from National Museums and Galleries of Wales for coordinating the trade show, to Amber Xavier-Rowe and Karen Dorn from English Heritage for marketing the conference and bringing the publication to press, and lastly to Peter Winsor from Resource for his support.

FOREWORD

This conference is a milestone in the development of integrated pest management (IPM) in museums, libraries, archives and historic houses. It brings together those people from many disciplines who are responsible for the care of collections and buildings and their preservation for the future.

In the past, pest control was a reaction to the discovery of insect activity and damage, with the response usually relying heavily on the use of toxic pesticides. The IPM approach is to look at the whole picture rather than to react to each crisis. The aims of IPM are to minimise use of pesticides and to provide practical, safe and cost-effective methods to prevent collections, furnishings and buildings from being damaged by pests.

All of the key IPM principles are covered in these conference papers. These include: identifying pests, detecting and monitoring pests, modifying the environment to discourage pest attack, developing safe control measures, and targeting treatment only where it is needed.

Care of collections and historic buildings involves many different disciplines. The expertise in collections care in museums, libraries and houses has been used to develop IPM programmes tailored to the specific needs of the collection or building. Experience has shown that IPM must be relevant to the needs of the collection and the building, whether historic or modern, using as much local information and expertise as possible. To succeed, it should also be a process of evolution rather than revolution and encourage participation by all staff.

Well-planned and well-executed IPM programmes have prevented problems occurring and prevented panic-driven crises. In times of reduced budgets, IPM programmes have been shown to make much more effective use of limited human and financial resources.

A vital part of the implementation of IPM is the training of staff and exchange of ideas. This conference enables participants from Canada to Singapore, with different collections, climates and pests, to come together and share knowledge for the benefit of all. We hope that these proceedings will stimulate others to adopt IPM and to share their knowledge and experience in the future.

Sir Neil Cossons
Chairman, English Heritage

Lindsay Sharp
Director, National Museum of Science and Industry

Lynne Brindley
Chief Executive, British Library

CHAPTER ONE

Museums, libraries and archives. The pests: their presence and the future

Robert Child
HEAD OF CONSERVATION

National Museums and Galleries of Wales, Cardiff CF10 3NP, Wales

Tel: +44 2920 573245   Fax: +44 2920 573124   e-mail: Robert.Child@nmgw.ac.uk

ABSTRACT

Insect control in museums has gone through a number of major technical and psychological changes since people started to collect, accumulate and preserve objects of historic and artistic importance. Since the advancement of science in the 19th century, strong and toxic poisons such as arsenic, mercuric chloride and cyanide were known to kill biological organisms. By the Second World War the first 'designer insecticides', such as DDT, had been developed to have low mammalian toxicity and environmental impact, but high insecticidal activity. These early specific insecticides were later replaced by 'safer' chemicals that did not have long-term, potentially damaging, residual potency. In the last decade, there has been a further reaction against the use of toxic insecticides on historical collections in favour of more refined methods of prevention, early detection and non-toxic control methods.

KEYWORDS

Insect pests, fumigation, insecticides, pest management

In 1957, the United Kingdom Working Party set up the Pesticides Safety Precautions Scheme, which monitored and advised on the toxicity and efficacy of pesticides. This voluntary scheme was eventually replaced by the statutory Control of Pesticides Regulations, 1986 (MAFF, 1986). Similar legislation exists in most developed countries.

The principle force behind regulation and legislation in the pest control industry was safety for the user and consumer, particularly in the agricultural and foodstuffs industries. However, in museums, libraries and other repositories of historical material, a different philosophy often existed; that of the long-term preservation of the artefacts. This was often achieved by large-scale prophylactic treatments that were intended to defend objects against insect attack.

Before the advent of modern scientific understanding in the 19th century, objects were collected for their inherent stability: metals, glass, ceramics, stone, etc. Even biological collections were based on materials that did not degrade easily, such as eggs, shells, horns, antlers, etc. Only taxidermy, based on the ancient techniques of skin curing, attempted to preserve organic-based materials, sometimes with little success, as the story of the last dodo demonstrates. The stuffed bird was displayed in the Pitt Rivers Museum in Oxford for only a few years before insect damage led to the body being destroyed, and now only the skull and feet survive.

The scarcity of much organic-based material from before the 19th century can be attributed to its destruction by fungal rots and insect pests. The predominance of oak furniture and house fittings prior to the 17th century in historic houses and museum collections led to this time being known as the 'age of oak'. It is now thought that many other woods such as beech, birch, ash, etc., were also used, but owing to their vulnerability to insect attack, these have largely been destroyed over the centuries by the ravages of woodworm. Insect attack was little studied and poorly understood, as this question posed to Lady Betty in 'Side Talks with Girls' in 'Home Chat' of 1897 shows:

It would be of great help to me if you would be good enough to tell me in your particular page, if it is possible to kill worms that have got into some oak furniture I bought. The chairs are modern ones, but have come from an old house. What have I to do to prevent further damage? Also, please tell me if it is a sign a bride is ready to receive callers when she appears in Church on Sunday?

Lady Betty's reply was 'Yes' to the last question. Now about the first one:

What you call worms is really dry rot ... a terrible infectious disorder which if not checked will probably cause the furniture to crumble to dust. You should soak the chairs thoroughly in paraffin – it is the only thing to destroy the insects.

The first commercial insecticides were produced in the early 20th century. Of particular importance was the Maxwell Lefroy formulation developed for treating Westminster Hall in London in 1912. It was eventually marketed by Lefroy in 1924 under the trade name of Rentokil (Ridout, 1998). Rentokil advertising literature of the period 'The Destroyers of your Furniture' promoted the fluid as 'its fumes (unlike the gases given off in fumigation) are not harmful to human beings or animals, and can be used with absolute safety'. The irony of this claim is that Professor Lefroy died of hydrogen cyanide poisoning and the Rentokil fluid, based on dichlorobenzene was, by 1950, described by the Forest Products Research laboratory as unsuitable for use inside buildings 'owing to the toxic effects of its fumes on man'. The principal concern of the early insecticidal treatments was an immediate and effective kill of pests. Thus many of the products had emotive names like 'Rentokil' and 'Hope Woodworm Destroyer'. The other essential factor was the effective life expectancy of the treatment, and early advertisements reflect this, as in Rentokil's claim 'Not one case of re-infestation since it was discovered 10 years ago'. Even in 1968, the Victoria and Albert Museum in its Technical Notes on the Care of Art Objects was stating:

> The success of insecticidal treatment is proportional to the penetration achieved. Hence woodworm-destroying fluids should be applied liberally to all absorbent surfaces and injected into worm holes and adventitious cavities.
> (Victoria and Albert Museum, 1968)

Together with commercial products, home made remedies and treatments remained popular especially with certain types of museum curator and antique dealer (Williams and Hawks, 1987). These remedies included the folk type such as this one printed in *Women's Own* in 1913:

> Mrs. Varley of Lancaster says 'red flannel has a strong attraction for moths, so after putting winter garments away, leave a square of flannel on the floor of the cupboard and the moths will eat this in preference to anything else'.

The Victoria and Albert Museum and other conservators were recommending do-it-yourself fumigation using such chemicals as carbon disulphide. Although their instructions made no mention of the extreme toxicity of this material, they do mention that it is highly inflammable 'and when mixed with air is explosive'. The liberal use of naphthalene and paradichlorobenzene has been widely recommended in conservation literature over the past 50 years as an invaluable aid in the prevention of insect attack (Goldberg, 1996).

The recognition that some insecticidal treatments could also damage materials by staining, chemical attack, etc., led to a number of alternatives being tried and recommended (Story, 1985). Regular microwaving of books has been proposed as a mass treatment method even though the

threat of burning from metal parts is mentioned (Brezner, 1988). More bizarre recommendations include the use of patrolling insect-eating spiders in libraries and archives. Non-webbing varieties such as the tarantula are thought suitable, but the author (Story) recognizes that 'defecation may still be a problem'. Whether this is caused by the spiders or alarmed visitors is not made clear.

Some home-based remedies have proved to be valuable and effective, although their value was not always immediately recognised. The British Columbia Museum's *Methods Manual* (Ward, 1976) confidently states 'beware those old remedies which suggest that either hot sunshine or freezing will destroy clothes moths and carpet beetles'. With the smugness of hindsight we now know that both high and low temperature treatments are routinely used in killing insect infestations in museum objects (Reagan, 1982).

It is interesting to note that with the establishment of the Pesticides Regulations Act, 1986, all insecticides had to be registered with the appropriate authorities. All manufacturers of pesticides have to obtain a certificate of approval for each pesticide product for use in the UK. A number of chemicals such as camphor, ethanol, ethyl acetate and thymol were approved for use as commodity chemicals, predominantly for use by some natural history curators and conservators.

There has been a welcome move away from the routine spraying of residual insecticides. In the past, museums and houses were dosed every few months with persistent toxicants whether there were insects present or not. Treatments are now much more targeted on specific problems and use insecticides, such as permethrin micro-emulsion, which are much more friendly towards staff, objects and the environment (Pinniger and Child, 1996).

The greatest changes in insect pest control over the last ten years have been in the large-scale treatments of objects. Formerly, these were routinely fumigated with toxic gases sometimes on an annual basis or more frequently (Forest Products Research Laboratory, 1950).

Fumigation, the killing of insects by the use of toxic gases, has unwittingly been used for centuries in food preservation where the build-up of carbon dioxide gas in sealed grain silos is an effective pesticide. In the 19th century, buildings were fumigated by burning sulphur, however, the noxious and highly acidic sulphur dioxide gas generated was not an efficient fumigant, and caused untold acid damage to furnishings and other materials. Later use of hydrogen cyanide was found to be cheap and effective with little residual odour or damage to objects. However, its extreme toxicity to humans limited its use (Berry, 1985).

By the 1940s, a number of fumigants were in common use including phosphine, hydrogen cyanide and ethylene oxide. A number of vapourizing liquids such as carbon tetrachloride, carbon disulphide and ethylene dichloride; and vapourizing solids such as thymol, naphthalene and paradichlorobenzene were also in use (Busvine, 1980).

While many museums continued to use these materials well into the 1980s and even to the present day, most have been discontinued for lack of efficacy, health and safety reasons and occasionally because of associated damage to the treated materials. In the post-war period, the most common fumigants used on historic collections were methyl bromide and, later in the US, sulphuryl fluoride (Anon., 1988). Ethylene oxide continued to be used for some years especially for large volumes of paper-based material owing to its fungicidal properties.

Methyl bromide attained popularity in many countries because of its speed and efficacy in killing most insect pests, its deep penetration into solid material such as wood, its rapid offgassing and its lack of chemical interaction with most materials. It can, however, react with some sulphur-based materials such as keratin and rubber, to give an unpleasant 'mecaptan-like' odour. It has recently been identified as an atmospheric ozone-depleter and its use worldwide is being phased out.

Museums have traditionally carried out their own small-scale fumigations using a number of chemicals (Linnie, 1987). Commonly used methods were vapourizing thymol crystals in a sealed cabinet to kill mould on affected books and paper (Byers, 1983), the use of naphthalene and paradichlorobenzene in insect cabinets to protect against insect attack, and fumigation using carbon disulphide vapour. In the 1970s, dichlorvos attained some considerable popularity in the UK and Europe because of its insecticidal vapour. A number of museums used PVC strips impregnated with liquid dichlorvos/DDVP to prevent insect attack in vulnerable areas such as natural history, ethnographical and textile collections. Even David Pinniger, then employed by the Ministry of Agriculture, Fisheries and Food, when investigating an outbreak of *Anthrenus verbasci* in Tring Museum, recommended 'that dichlorvos slow-release units be installed to protect specimens in the public galleries and bird collection from re-infestation' (Scudamore *et al.,* 1980). The swing away from the use of any form of toxic treatments in the last few years has been such that recommendations like this would be unacceptable today – only 20 years after they were originally made. However, it must be said that the use of DDVP slow-release strips at the correct dose rates at Tring and the Natural History Museum main site have completely prevented any damage to stored or displayed animal specimens over this period. It seems likely that the use of DDVP slow-release strips will be restricted in the near future. As there is no effective replacement vapour insecticide, museums will have to invest in much better detection and monitoring systems and greatly improved cleaning regimes. Otherwise, they will suffer from increased damage to their previously DDVP-protected collections.

The last 10 years has seen the development by museums, libraries and those in charge of historic collections, of treatments specifically for heritage material.

These treatments meet the criteria demanded by conservators and curators in terms of operator safety, third party liability, efficacy, and are non-residual and non-damaging to the objects being treated (Stansfield, 1989). Anoxic fumigation with gases such as nitrogen and argon are now routinely carried out in many countries, and the use of low-temperature and high-temperature treatments under rigorous controls are commonplace. There has even been the development of low-hazard insecticides specifically for museum collections, which have met conservation testing standards. Perhaps, after a long and convoluted evolutionary process, museum insect pest control has now come of age. Perhaps we, as the custodians of much of this world's heritage, can be satisfied with one aspect of its preservation.

REFERENCES

Anon., 'Vikance holds potential as a museum fumigant', *Getty Conservation Newsletter,* Winter 1988, 6.

Berry R W, 'Recent developments in the remedial treatment of wood-boring insect infestations', in *Biodeterioration,* 1985, **5**, 154–165.

Brezner J, 'Protecting books from living pests', in *Proceedings of 1988 Paper Preservation Symposium,* 1988, Washington DC, USA.

Busvine J R, *Insects and Hygiene.* Chapman and Hall, London, 1980.

Byers B, 'A simple and practical fumigation system', *The Abby Newsletter,* 1983, **7**, 43 supplement 1.

Forest Products Research Laboratory, 'The common furniture beetle', *Leaflet No. 8 Crown copyright London,* 1950.

Goldberg L, 'A history of pest control measures in the anthropology collections of the National Museum of Natural History, Smithsonian Institution', in *Journal of the American Institute of Conservation,* 1996, **35**, 23–45.

Linnie M J, 'Pest control: A survey of natural history museums in Great Britain and Ireland', in *The International Journal of Museum Management and Curatorship,* 1987, **6**, 277–290.

MAFF, *Control of Pesticides Regulations 1986.* Food and Environmental Protection Act 1985, Ministry of Agriculture, Fisheries and Food, 1986.

Pinniger D B, Child R E, 'Insecticides: optimising their performance and targeting their use in museums', in *Proceedings of the 3rd International Conference on Biodeterioration in Cultural Property,* Bangkok, 1996.

Reagan B M, 'Eradication of insects from wool textiles', in *Journal of the American Institute of Conservation,* 1982, **21**, 1–34.

Ridout B, 'A history of deathwatch beetle, research and treatment', in *Conference Abstract of Wood Care Conference English Heritage London,* 1998.

Scudamore K A, Pinniger D B, Hann J J, 'Study of dichlorvos slow-release units used in museum cases for the protection of specimens against insect infestations',

Tring Museum 1978–79. MAFF Agricultural Science Service Pest Control Chemistry Department Report No. 45, 1980, London.

Stansfield G, 'Physical methods of pest control', in *Journal of Biological Curation,* 1989, **1**(1), 1–3.

Story K O, 'Approaches to pest management in museums', *Conservation Analytical Laboratory,* 1985, Smithsonian Institution, Maryland, USA.

Victoria and Albert Museum, 'Woodworm in furniture', *Technical Notes on the Care of Art Objects,* 1968, **2**.

Ward P R, 'Getting the bugs out', *Museum Methods Manual,* 1976, British Columbia Provincial Museum.

Williams S L, Hawks A H, 'History of preparation materials used for recent mammal specimens', in *Mammal Collection Management,* 1987, **IV**, 21–29.

BIOGRAPHY

Robert Child trained as a chemist and was originally a research chemist in the oil industry. He is currently the Head of Conservation at the National Museums and Galleries of Wales and acts as advisor on Insect Pest Control to the National Trust. He is a consultant on preventative conservation, especially pest management.

CHAPTER TWO

INSECT PESTS IN HISTORIC BUILDINGS: MISUNDERSTOOD, MISDIAGNOSED AND MISTREATED

Dr Jagjit Singh

DIRECTOR OF ENVIRONMENTAL BUILDING SOLUTIONS LTD

30 Kirby Road, Dunstable LU6 3JH, Bedfordshire, United Kingdom

Tel: +44 1582 690187   Fax: +44 1582 690188   e-mail: ebs@ebssurvey.co.uk

Website: http://www.ebssurvey.co.uk

ABSTRACT

Insect pests have evolved to exploit the man-made spatial ecosystems of historic buildings and microclimates of historic collections. A range of insect pests is encountered in our buildings, where they exploit different ecological niches and feed on a variety of substrates. A thorough understanding of the biology of the pest organism, together with its correct detection and diagnosis, should form an essential part of the measures taken to control it. Many current control practices rely on treating the symptoms, rather than dealing with the fundamental causes. This approach is neither sympathetic to the historic building fabric or to the collections it houses, nor does it provide a long-term solution to the problem. In the last century, the management of insect pests has largely relied on misunderstanding, misdiagnosis and mistreatment, causing considerable damage to the collections, historic fabric and building occupants. This paper will emphasize the interrelationship of the built and indoor environment with living organisms, organic materials and historic collections. It will focus on the benefits of successful environmental manipulation and control, rather than remedial 'belt and braces' treatments.

The key issues of this paper are:

- non-destructive, diagnostic environmental inspection for insect pests
- monitoring masonry moisture profiles
- monitoring timber moisture profiles
- monitoring and mapping the indoor environment
- analysis and interpretation of data
- developing holistic, tailor-made solutions
- correct identification and diagnosis of insect pests
- extracts from recent case studies
- health effects of the unnecessary treatments

This paper will present a multidisciplinary, integrated approach for providing holistic, sustainable solutions to the management of insect pests in collections and historic buildings.

KEYWORDS

Non-destructive inspection, environmental control, environmental monitoring, sustainable solutions, multidisciplinary approach to integrated pest management (IPM)

INTRODUCTION

Over the long history of human evolution, various insects and pests have evolved to exploit the man-made indoor and built environment. The study of insect pests associated with the built environment has increased in importance, as society becomes more aware of their aesthetic, economic and medical impact on the quality of life (Mukerji *et al.*, 2000; Singh and Aneja, 1999; Singh and Walker, 1996).

Providing an acceptable quality of life, including a pest-free indoor environment, will be an important issue in this century. However, increased public awareness and pressure from the environmentalist lobbies, and environmental and medical concerns about the use of insecticides and pesticides in the living and workplace environment, dictate radical and more progressive strategies for pest control (Robinson, 1996).

An understanding of the physical, biological and chemical control methods and their integration with mechanical and cultural factors will enable us to develop more targeted treatment, rather than the current practices of blanket 'belt and braces' treatment. The integrated insect pest management (IPM) programme will be the way forward for the future pest control in the indoor and built environment.

NON-DESTRUCTIVE, DIAGNOSTIC ENVIRONMENTAL INSPECTION FOR INSECT PESTS

Just as physicians and surgeons use stethoscopes and keyhole surgery to examine internal organs and to check their patients, building pathologists can use a range of non-destructive instrumentation to check their buildings. This will save unnecessary exposure work and the destruction of the historic fabric (Singh, 1991a).

Just as pathology is the scientific study of the cause and effects of disease, building pathology similarly encompasses not only observation of the structural and functional changes of the performance of the building, but the elucidation of the factors which cause it. For example, the cure for athlete's foot is not amputation, but drying out of the affected parts and localized targeted treatment with fungicide if necessary.

Non-destructive inspection techniques enable the condition of materials and contents of the buildings and collections to be ascertained without opening much of the building fabric; they are therefore especially valuable in buildings of historic and architectural interest (Singh, 1991b).

DIAGNOSTIC, ENVIRONMENTAL INSPECTION

The diagnostic approach involves carrying out regular inspections using a range of instrumentation and non-destructive investigation techniques. This approach enables a specific maintenance programme to be drawn up and ensures that the loss of historic fabric is kept to a minimum (Singh, 1994).

The total reliance on one method of inspection or isolation technique for pest infestation organisms is not ideal. A combination of the use of instrumentation with common sense is the best way forward. For example a number of *in situ* methods for decay assessment are available; however, these involve destructive techniques and have had varying degrees of success in detecting decay and predicting the residual strength of structural materials.

The analysis of decay organisms with destructive sampling is neither sympathetic nor acceptable to the conservation of historic churches, castles, abbeys, monuments and other landmarks.

MONITORING AND MAPPING THE INDOOR ENVIRONMENT

Historic building materials, collections and contents are inherently susceptible to fungal and insect infestation and decay, if not kept dry and well ventilated. Once the infestation has started it will continue to propagate, if the conditions are favourable, until eventually the materials can no longer sustain loads. It is therefore important that the building or structure is regularly evaluated for decay to prevent failure or collapse (Singh and White, 1995).

Based on this information, environmental control measures can be put in place to prevent further advancement of the decay.

ASSESSMENT OF DECAY ORGANISMS

Assessment of the activity of decay organisms involves the following:

- detection of decay organisms
- identification of decay organisms
- assessment of the viability of decay organisms
- quantification of the state of decay
- environmental conditions assessment
- structural assessment of decay

It is common for decay organisms to be present and actively causing decay without any apparent external manifestation. In these cases their detection presents a considerable challenge, and there are currently no completely satisfactory methods available. A number of non-destructive techniques employing high-technology equipment have been found to be useful for advanced stages of decay, but earlier stages are often difficult to detect. For example, the use of a resistograph to assess the condition of timber for insect and fungal infestation and decay in timber statues and artefacts, timber member in buildings (see Figures 1, 2 and 3) and living trees.

Figure 1 The resistograph in use to assess the condition of timbers

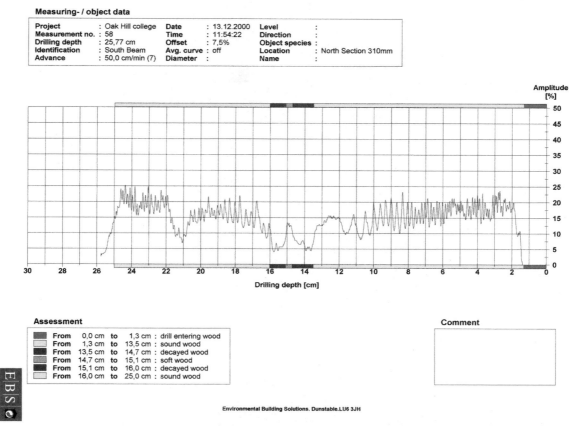

Figure 2 Non-destructive inspection of timber: a graph from resistograph drilling in the north section of Oak Hill College

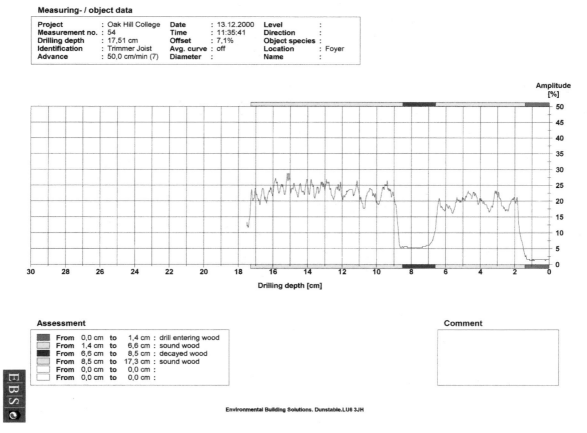

Figure 3 Non-destructive inspection of timber: a graph from resistograph drilling in the foyer of Oak Hill College

The most obvious signs of infestation are the presence of insect pests and associated excretory products, together with the distinctive damage they can cause, for example, woodboring insects in timbers. However, these indicators are often the latest signs of decay, by which time the majority of the damage has already been done (Drdacky *et al.*, 1994).

THE CONSERVATION APPROACH: DEVELOPING HOLISTIC, TAILOR-MADE SOLUTIONS

An initial assessment of the symptoms and causes of the problems in both modern and historic buildings can be made by the use of commonly available equipment, such as binoculars, small mirror and/or surveyor mirror, screwdriver, ladder, torches, compass, camera and moisture meters.

Building material in historic buildings, churches and other structures is inherently susceptible to infestation and decay, if it is not kept dry and well ventilated. Once the infestation has started it will continue to propagate, if the conditions are favourable, until eventually the material can no longer sustain loads.

Remedial chemical treatments for woodworm, death-watch beetles, dry rot and wet rots are very expensive and often cause more damage to the health of building fabric and people than the infestation itself. It is therefore important that the building or structure is regularly evaluated for material decay and assessment to prevent failure or collapse, which could result in loss of use. Based on this information, environmental control measures can be put in place to prevent further advancement of the decay.

These methods provide the most cost-effective, long-term, holistic and environmentally sustainable conservation solution for the restoration of timber structures, including bridges, jetties, transmission poles, historic buildings and monuments.

STABILIZING THE HISTORIC ENVIRONMENT

Once the above investigations have been carried out, a strategy can be put forward to stabilize the building environment.

It is important to stabilize the historic building environment. For the holistic and sustainable conservation and preservation of the building, various building works will be required to prevent further water penetration and to maximize ventilation to damp affected materials. Correction of these building defects, combined with measures to dry out the wet areas and to protect the decorative interior finishes by allowing ventilation of the wet areas, will prevent further deterioration.

Until the drying out of the building fabric and its associated timber elements is completed, any other actions to remedy the deterioration problems will be ineffective and a waste of time and resources. Continuous long-term monitoring and preventative maintenance of the building may be necessary:

- to assess the state of moisture equilibrium and balance (moisture sources, reservoirs and sinks) in the building environment, building fabric and structural elements as the building dries out
- to allow co-ordination and scheduling of work stages to prioritize the remedial work to achieve acceptable levels of moisture in the masonry and timber to prevent future deterioration problems
- to allow a cost-effective, long-term holistic approach to environmental stabilization of the historic environment

REFERENCES

Drdacky M, Palfreyman J W, Singh J, *Conservation and Preservation of Timber in Buildings,* 1994, Aristocrat, Telc, Czech Republic.

Mukerji K G, Chamola B P, Singh J, *Mycorrhizal Biology,* 2000, Kluwer Academic/Plenum Publishers, USA.

Robinson W H, *Urban Entomology, Insect and Mite Pests in the Human Environment,* 1996, Chapman and Hall, London.

Singh J, 'Non-destructive investigation', in *Building Research Information,* 1991a, **19**, 20.

Singh J, 'New advances in identification of fungal damage in buildings', in *Mycologist,* 1991b, **5**, 139–140.

Singh J, *Building Mycology, Management of Health and Decay in Buildings,* 1994, Spon, London.

Singh J, Aneja K R, *From Ethnomycology to Fungal Biotechnology, Exploiting Fungi from Natural Resources for Novel Products,* 1999, Kluwer Academic/Plenum Publishers, USA.

Singh J, Walker W, *Allergy Problems in Buildings,* 1996, Mark Allen Publishing Ltd, UK.

Singh J, White N, *Environmental Preservation of Timber in Buildings,* 1995, Oscar Faber, St. Albans, UK.

BIOGRAPHY

Dr Jagjit Singh, Director of Environmental Building Solutions Ltd., is an independent consultant specializing in building health problems, heritage conservation and environmental issues. His current research focuses on interrelationships of building structures, materials and collections with their environments and occupants. Dr Singh appeared as the country's leading expert on damp and woodworm in BBC2's series 'Raising the Roof'.

CHAPTER THREE

NEW PESTS FOR OLD: THE CHANGING STATUS OF MUSEUM INSECT PESTS IN THE UK

David Pinniger

CONSULTANT ENTOMOLOGIST

83 Westwood Green, Cookham, Berks SL6 9DE, United Kingdom
Tel: +44 16285 26066   Fax: +44 16285 23233   e-mail: david@pinniger.globalnet.co.uk

Illustrations by Annette Townsend, Conservation Officer
National Museums and Galleries of Wales, Cathays Park, Cardiff CF1, Wales

ABSTRACT

Some pests such as the common clothes moth *Tineola bisselliella* and varied carpet beetle *Anthrenus verbasci* are known to have been in the UK for many years. Two beetle species, which were unknown in the UK before 1960, have now become established in museums and cause serious damage to objects. The Guernsey carpet beetle *Anthrenus sarnicus,* was first recorded in London in 1963 and is now widespread in west London and parts of south-east England. It has also spread to museums in Cambridge, Edinburgh and Liverpool. The brown carpet beetle (or vodka beetle) *Attagenus smirnovi* arrived in London in the 1970s and appears to be spreading, although at a slower rate than the Guernsey carpet beetle. Three other Dermestid beetles – *Reesa vespulae, Thylodrias contractus* and *Trogoderma angustum* – have now become established in museums in the UK. Although at the moment their range is restricted, all have the potential to cause serious damage to natural history collections. Vigilance in identifying and distinguishing these new pests, coupled with effective inspection and quarantine procedures, should help reduce the risk of increased damage to collections.

KEYWORDS

Museum pests, carpet beetles, *Anthrenus, Attagenus, Reesa, Thylodrias, Trogoderma*

INTRODUCTION

Insect pests were not always pests and we need to look for natural habitats to see where they originated. Most of our textile pests, such as clothes moths and carpet beetles, occupy an environmental niche in bird and mammal nests. Here they perform a very important role in the breakdown and recycling of hair, skin, feathers and excreta. Pests of stored food, such as grain weevils and biscuit beetles, can be found in caches of food, which some animals store over the autumn and winter. Some pests have been associated with humans for many years. Clothes moths caused problems for the Ancient Greeks and Romans and are mentioned in the Bible. Woodborers were probably stowaways on Noah's Ark and woodworm, *Anobium punctatum,* has been in the UK at least since Roman times. The Ancient Egyptians also had problems with pests, and the biscuit beetle *Stegobium paniceum,* cigarette beetle *Lasioderma serricorne* and hide beetle (*Dermestes* species) have all been found associated with entombed burial materials. It is clear that international trade has played an important part in the spread of many pests. The movement of large quantities of food by ship in the 19th century

resulted in many pests such as biscuit beetles, flour beetles and spider beetles reaching a large number of countries. Trade in other commodities such as wool and timber helped to spread carpet beetles, clothes moths and woodborers.

Over the years, the change in emphasis has been from the damage to personal belongings to include collections of objects specifically preserved for posterity. In addition, the pest status of many insect species has changed over time. Although clothes moth problems seem to have increased in recent years, there has been a marked decline in active infestations of woodworm. Other species such as the golden spider beetle *Niptus hololeucus,* have been known in the UK for many years, but only recently have become common in historic houses. Many of these changes in pest status can be linked directly to changes in environmental conditions and in storage and display methods. There is now also much greater opportunity for insects to move from one country to another, due to ease of travel, which encourages much greater exchange of material between collections.

Despite the spread of pests, there are some distinct differences between pests found in different climatic areas. For example, the distribution of drywood and subterranean termites can be directly linked to the climatic conditions found in tropical and subtropical countries. The distribution of other pests may be much more local. For example, there is a marked decrease in the frequency of carpet beetles in the UK from south-eastern England to northern England, and they are relatively rare in Scotland. Evidence from pest records in recent years shows that changes in pest status are occurring more rapidly and some pests, which were previously unknown in the UK, have become established. Halstead (1975) warned of the pest potential of some species that had been recently introduced into the UK. He was proven absolutely right, as two species of carpet beetles (*Anthrenus sarnicus and Attagenus smirnovi*) are now established in London and cause serious damage to collections. I have been closely involved with monitoring the recent spread of these two species. Their origin, biology and pest status will be described, together with three other species with serious pest potential which have been recently been recorded in UK museums.

GUERNSEY CARPET BEETLE '*ANTHRENUS SARNICUS*' (MROCZKOWSKI)

This species was first described in 1963 from specimens found in a house in Guernsey in 1961. It was first found in South Kensington in London in 1963 (Peacock, 1993). It is very similar in appearance to the common and well-known varied carpet beetle *Anthrenus verbasci*. The adults are 3–4 mm long and paler in colour than *Anthrenus verbasci* with an overall silvery-grey appearance (Figure 1). The main diagnostic feature is that the scales of *Anthrenus sarnicus* are triangular. This is well illustrated in Peacock (1993) but the shape can only be seen under a microscope. The larvae are very similar to those of *Anthrenus verbasci* and although *Anthrenus sarnicus* tend to be more gingery-brown and paler in colour, they can be very difficult to distinguish.

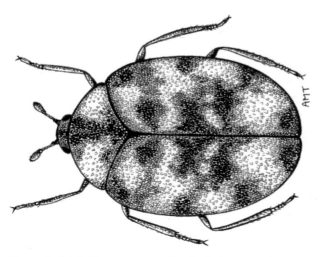

Figure 1 Adult Guernsey carpet beetle *Anthrenus sarnicus*

A study by Armes (1988) confirmed that *Anthrenus sarnicus* was established in the Natural History Museum and that it had spread to bird nests near the museum. It has since become the major pest in this museum and has substantially replaced the varied carpet beetle *Anthrenus verbasci*, previously the dominant pest. *Anthrenus sarnicus* soon spread to the adjacent Science Museum and to the Victoria and Albert Museum, 100 m away. Movement of objects from both the Victoria and Albert Museum and the Science Museum to the main store at Blythe House in Olympia enabled *Anthrenus sarnicus* to travel west and infest the store. Other early recordings were from the Commonwealth Museum, and the first infestation in a museum outside the London area was in Liverpool. The larvae of *Anthrenus sarnicus* will attack a wide range of materials. They have severely damaged natural history specimens including insects, crustacea, dried animal skins, baleen, feathers and fur. They have also damaged many textiles, particularly wool, felt, tweed and jersey.

The spread of *Anthrenus sarnicus* has continued and, although it appears to be more common in south-east England, it has also been found in Salisbury, Cambridge, Barnard Castle, Chatsworth and Edinburgh. The discovery of the sex attractant pheromone by Finnegan and Chambers (1993) enabled us to bait traps with lures of decyl butyrate and decanol. These were used in the Natural History Museum in London and in the Natural History section of the Liverpool Museum (Ackery *et al.*, 1999). The lures also helped to detect this species in other places, such as the British Museum in London and at Oxford, Newbury and Romsey. I have recently found *Anthrenus sarnicus* damaging textiles in a number of private houses in London and near Watlington, Oxon, and this species should now be regarded as a domestic pest of some importance. It has not so far been recorded in any other countries but as it is easily confused with other *Anthrenus* species and is rarely mentioned in references, it may have been overlooked.

It is clear that *Anthrenus sarnicus* has been very successful in exploiting the museum and domestic niche and replacing *Anthrenus verbasci*. There are many reasons which influence the success of a pest and those which have given *Anthrenus sarnicus* the edge include faster development with no diapause, a tolerance of higher temperatures, good dispersal with very mobile larvae and active, strong-flying adults. With the trend to warmer winters in the UK, it is likely that this species will become far more common as a domestic pest and will spread more widely into museum collections.

BROWN CARPET BEETLE (VODKA BEETLE) '*ATTAGENUS SMIRNOVI*' (ZHANTIEV)

This species was first found in Moscow in 1961 and was first described in 1973. In 1978, specimens were found in a flat in Kensington in London (Peacock, 1979). It was first

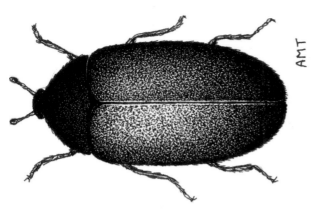

Figure 2 Adult brown carpet beetle (Vodka beetle) *Attagenus smirnovi*

found in the Science Museum in 1989 where a serious infestation was eventually discovered in cattle food in a milking parlour display. Soon after, adults and larvae were found on sticky traps in the Victoria and Albert Museum. The name *Attagenus smirnovi* led to it being called the vodka beetle, an adopted name, which has stuck. The adults are similar in shape and size (5–8 mm) to the very common two-spot carpet beetle *Attagenus pellio,* but *Attagenus smirnovi* have brown elytra with no spots (Figure 2). The larvae are also very similar to the larvae of *Attagenus pellio* and can only be distinguished by the arrangement of body scales (Peacock, 1993).

Attagenus smirnovi is well known in Russia and Eastern Europe and is now a major domestic pest in Denmark. There are now established populations in all three national museums in South Kensington and damage has been recorded on some textiles in the Victoria and Albert Museum. Although it is widespread in the Natural History Museum and is regularly found on traps in galleries and stores, there is as yet little evidence of it causing damage to specimens. Infestations in the Science Museum have been restricted to starchy-based materials such as grain and cattle food. *Attagenus smirnovi* has been carried on objects to the Blythe House store in Olympia and is now established in the building. Although it is widespread and present on all four floors, the only really serious infestation was associated

with food spillage and poor hygiene in mess areas (Kingsley and Pinniger, 2001). The larvae are mobile and the adults are active fliers attracted to tungsten lights. Unlike the Guernsey carpet beetle, this pest has not spread widely in the UK and, so far, the only other places outside London where it has been found are Cambridge and Chatham. It may be that the more specific dietary requirements of this species restrict its success and its spread. It is rarely mentioned in pest literature and adults have been misidentified as faded specimens of the black carpet beetle *Attagenus unicolor.* Wider awareness of its occurrence in the UK may increase the number of records and evidence of damage by this species.

MUSEUM NUISANCE/AMERICAN WASP BEETLE 'REESA VESPULAE' (MILLIRON)

This insect does not have a recognized common name and is very unusual in that it is parthenogenetic. This means that all the beetles are females and can lay fertile eggs without mating. As there are no male insects and one insect can start an infestation, the pest potential is greatly increased. *Reesa* is found in northern Europe and the USA where its natural home is insect nests. It has now become a major domestic pest in parts of Scandinavia. The adults are easily recognized as they are about 2–4 mm long and have a yellowish band of hairs across the elytra (Figure 3). The larvae are tubular and taper towards the rear, which has a tuft of hairs. They can easily be mistaken for *Trogoderma* species.

Reesa was first recorded in the UK in 1977 from seeds in Essex (Adams, 1978). It was first found in the Natural History Museum in London in 1979 and is now well established in the entomology department where it regularly causes persistent and serious damage to specimens in isolated drawers. Fortunately, it has not spread widely in the museum and does not seem to have caused damage to other types of natural history specimens. The larvae are known to feed on dried insects but they will also attack seeds and freeze-dried animals. Apart from the Natural History Museum and an occurrence in the National Museum of Scotland in Edinburgh, there are no other published records of infestations in UK museums. But in view of its persistence, other museums should be very vigilant to prevent this pest becoming established.

CABINET BEETLE 'TROGODERMA ANGUSTUM' (SOLIER)

This is the most recent pest introduced to the UK and does not yet have a common name. It was first recorded by Shaw (1999) in the National Museum of Scotland in Edinburgh in 1998, where it was found damaging bird specimens. It was also established at this time in the herbarium of the Royal Botanic Gardens in Edinburgh, but the larvae may have been misidentified as *Reesa vespulae* at an earlier date, as they are very similar in appearance.

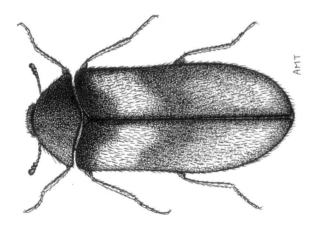

Figure 3 Adult museum nuisance beetle (American wasp beetle) *Reesa vespulae*

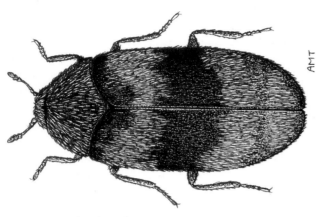

Figure 4 Adult cabinet beetle *Trogoderma angustum*

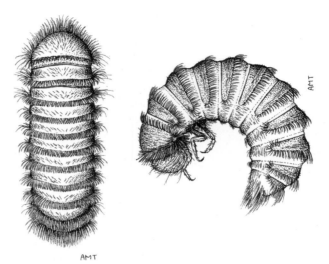

Figure 6 Larva of odd beetle *Thylodrias contractus:* (a) side view, (b) top view

The adults are 2–4 mm long and are quite distinctive with two or three bands of whitish hairs on the elytra (Figure 4).

It has a very restricted occurrence in other countries, but is a major pest in the Natural History Museum in Stockholm and appears to be coming more widespread in Europe (Akerlund, pers. comm.). The first record from a museum in England was a specimen found in the herbarium of the Royal Botanic Gardens in Kew in 2000. Subsequently, larvae found in the herbarium thought to be *Reesa vespulae* were reared into adult *Trogoderma angustum*. Museum staff should be aware of the potential of this pest to cause serious damage to plant and animal specimens.

ODD BEETLE 'THYLODRIAS CONTRACTUS' (MOTSCHULSKY)

This is called the odd beetle, and rightly so, as the male has wing cases that remain partly open and is a very strange shape for a Dermestid beetle (Figure 5). The female has no wing cases or wings and looks like a louse (Peacock, 1993). The larvae are also very distinctive as they are short and fat with hairy bands and when disturbed, they curl up like a woodlouse (Figures 6a and 6b). *Thylodrias* is native to central Asia and has become established in some museums such as the Smithsonian in Washington. It has also become established as a pest in museums and houses in Finland

where the larvae are known to eat dead insects and attack natural history specimens. The female is flightless and has a pheromone to attract males.

The only known established breeding population of the odd beetle in the UK is in the herbarium in the Natural History Museum in London. It was first found in 1984 and small numbers are found each year on sticky monitoring traps. There are no records of it attacking plant specimens and the larvae may be living on dead insect debris. Fortunately, it has not yet been found in the entomology department. The only other place where I have found this species is at West Dean College in West Sussex. Two larvae were found on sticky traps placed in a fireplace as a survey exercise in a jointly run Getty Conservation Institute and Museums and Galleries Commission (MGC) pest management training course, held in 1996. There was known to be an old bees' nest in the chimney. An interesting link between West Dean College and the USA is that the late owner, Edward James, had large estates in Virginia and regularly moved objects between the USA and the UK.

CONCLUSIONS: WHAT OF THE FUTURE?

There is no doubt that the Guernsey carpet beetle is now an established museum and domestic pest in the UK. It is now more important than the varied carpet beetle in some areas and, in 30 years, it has spread from its bridgehead in South Kensington. How far the spread of *Anthrenus sarnicus* will continue will depend upon opportunity and the restrictions of climate.

There are other pest species which have established a foothold in the UK. The cigarette beetle *Lasioderma serricorne* has recently been found in the herbarium at Kew. Breeding populations of this species are not normally found in the UK but it is the most serious pest of herbaria in hot countries. The first established termite colony (of *Reticulitermes*) was recently found in Devon (Anon, 1999).

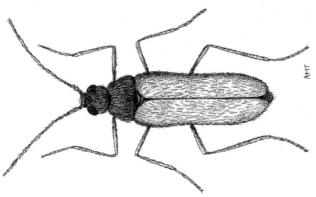

Figure 5 Adult male odd beetle *Thylodrias contractus*

These most feared of insects, have now been allegedly eradicated, but threats of global warming were given a new perspective by their discovery in England.

Termites and other recently introduced pests have the potential to become serious pests in the UK depending upon opportunity and changing climate. It is essential that staff are aware of these new species and can identify them so that new occurrences are not overlooked. This can be a problem when new pests do not appear in the literature and have not yet received a recognized common name. Pest monitoring coupled to quarantine and treatment of new or returned material are the key components of pest prevention. Early warning of pests, such as *Reesa*, may enable eradication of a local population before it becomes established and causes serious damage.

Obscure insects are not only of interest to eccentric entomologists but may also have the potential to be the pests of the future.

REFERENCES

Ackery P R, Chambers J, Pinniger D B, 'Enhanced pest capture rates using pheromone-baited sticky traps in museum stores', in *Studies in Conservation,* 1999, **44**, 67–71.

Adams R G, 'The first British infestation of *Reesa vespulae*', in *Entomologists Gazette,* 1978, **29**, 73–75.

Anon., 'Termites and UK buildings: biology, detection and diagnosis', in *BRE Digest*, 1999, **443**(1), 12.

Armes N J, 'The seasonal activity of *Anthrenus sarnicus* and some other beetle pests in the museum environment', in *Journal of Stored Product Research*, 1988, **24**, 29–37.

Finnegan D E, Chambers J, 'Identification of the sex pheromone of the Guernsey carpet beetle *Anthrenus sarnicus*', in *Journal of Chemical Ecology*, 1993, **19**, 971–982.

Halstead D G H, 'Changes in the status of insect pests in storage and domestic habitats', in *Proceedings of the 1st International Conference on Stored Product Entomology,* 1975, Savannah, USA, 142–153.

Kingsley H and Pinniger D B, 'Trapping used in a large store to target cleaning and treatment, in *Integrated Pest Management for Collections, Proceedings of 2001: A Pest Odyssey* (this volume), James & James (Science Publishers) Ltd, London, 51–56.

Peacock E R, '*Attagenus smirnovi*, a species new to Britain with keys to adults and larvae of British *Attagenus*', in *Entomologist's Gazette*, 1979, **30**, 131–136.

Peacock E R, 'Adults and larvae of hide, larder and carpet beetles and their relatives', in *Handbooks for the Identification of British Insects*, 1993, **5**(3), Royal Entomological Society, London.

Shaw M R, '*Trogoderma angustum*, a museum and herbarium pest new to Britain', in *Entomologist's Gazette*, **50**, 99–102.

ACKNOWLEDGEMENTS

Thanks to all my friends and colleagues in museums who have checked traps and provided information and insect specimens. Additional thanks go to Mark Shaw and David Halstead for helpful comments on the taxonomy and pest histories.

BIOGRAPHIES

David Pinniger is an entomologist who provides consultancy and training on pest management. He was a research entomologist at the MAFF Central Science Laboratory but is now an independent consultant. He is the pest management strategy adviser for English Heritage and many of the major UK museums such as the Victoria and Albert Museum and the Science Museum.

Annette Townsend is a conservation officer at the National Museums and Galleries of Wales in Cardiff. She is a conservator of wax models and illustrator for the Department of Biodiversity.

CHAPTER FOUR

Understanding and controlling anobiid beetles with special reference to the deathwatch beetle *Xestobium rufovillosum*

Brian V Ridout

Ridout Associates, 147a Worcester Road, Hagley, Stourbridge, West Midlands, DY9 0NW, United Kingdom
Tel: +44 1562 885135   Fax: +44 1562 885312   e-mail: ridout-associate@lineone.net

Abstract

In 1991, Ridout Associates were commissioned by English Heritage to make a detailed study of timber decay in English cathedrals. This task required an appraisal of all published information on the deathwatch beetle and furniture beetle. It soon revealed that previous research had been biased towards laboratory experimentation, and that the insect's natural behaviour and requirements in trees and buildings had been largely ignored. The practical consequence was that huge volumes of insecticides were applied to timbers that could never be attacked by the beetles, or in a situation where infestation was historical and extinct.

Subsequent research, funded by English Heritage (1991–1994), the European Commission DGX11 Programme (1995–1997) and the DTI (2000), has provided us with a greater understanding of insect damage potential. This paper will provide information on ecology-based and behaviour-based trapping techniques for the deathwatch beetle and furniture beetle. It will then show how a greater understanding of these topics enables us to avoid unnecessary treatments, and to formulate an integrated approach for beetle control.

Keywords

Anobiid beetle, infestation, integrated approach

Introduction

The Coleopteran family *Anobiidae* contains many species of woodboring beetles, which live in the dead parts of standing trees. Some of these beetles will invade and thrive in buildings and artefacts if environmental conditions are suitable. These insects, popularly called furniture beetles, deathwatch beetles, powderpost beetles or woodworm, cause significant damage throughout the more temperate regions of the world.

Insect damage in timber has usually been addressed by the surface application of insecticides. However, in many cases, the insects are no longer present, much of the timber is naturally resistant to attack, or physical problems, e.g. finishes, timber size or accessibility make eradication uncertain. This paper examines an integrated approach to treatment, with special reference to the deathwatch beetle. This approach was researched and devised during an international programme of investigation, which was initiated by English Heritage and part financed by the European Environment Programme (DGX11). The aim was to understand the behaviour of deathwatch beetles in buildings, and to assimilate these findings into a scientific approach for beetle control. Recent research, part funded by the Department of Trade and Industry (DTI), has

expanded this research to include *Anobium punctatum*, the common furniture beetle.

Woodboring anobiid beetles in buildings: a brief world review

Many species of woodboring anobiid beetles occur in forests around the world, but only a few are of any significance in buildings. Thus 332 species of *Anobiidae* in 52 genera have been recorded from the USA (USDA, 1982), but most damage is caused by *Hadrobregmus carinatus* and *Ptilinus ruficornis* in the north-east (Simeone, 1961) and *Hemicoelus gibbicollis* in the north-west (Suomi and Akre, 1993). Nevertheless, the damage caused by these insects is significant. Suomi and Akre (1993) found *Hemicoelus gibbicollis* in nearly all of the 90 Washington State homes and outbuildings they investigated during 1987–91, whilst Mauldin *et al.* (1986) estimated the cost of treatment and repairs necessitated by woodboring beetles as $50,000,000 per year in the USA.

Elsewhere in the world the most widespread and significant anobiid beetle in buildings is probably *Anobium punctatum*. This insect occurs throughout Europe, and was introduced into Australia (Peters and Fitzgeralds, 1996), New Zealand (Hosking, 1978) and the USA (Simeone, 1961). The distribution of *Anobium punctatum* in Europe overlaps to some extent that of *Nicobium* spp. and *Oligomerus ptilinoides*, two highly destructive genera. *Anobium punctatum* predominates in the north, whilst *Nicobium* and *Oligomerus ptilinoides* become major pests of buildings in southern Europe and the Mediterranean regions (Halperin, 1992; Nour, 1962).

These beetles can mostly tolerate quite dry conditions (Table 1) and their activities tend to be restricted to sapwood, or the heartwood of perishable species. Some anobiid beetles, however, thrive in wetter conditions and are able to attack the heartwood of durable timbers if the wood chemistry has been modified by fungus. The most significant species in this group is the deathwatch beetle *Xestobium rufovillosum*. This insect is a serious pest in the medieval oak timbers of Western Europe. It also strays into the warmer climates of northern Spain (Espanol, 1969), Italy (Veronese, 1995) and Greece (Ridout, unpublished data). *Xestobium rufovillosum* was recorded in 1937 from the Old South Meeting House in Boston, USA (Muirhead, 1941) and the beetle is now found quite frequently in the Connecticut area, where it was presumably introduced from ships timbers or furniture. The climate seems unlikely to provide a barrier to a significant increase in distribution, and it may be that the major limiting factor is the historical use of suitable hardwoods in construction.

Two other significant woodboring anobiids, which apparently require damp timber and decay, are *Hadrobregmus pertinax* and *Priobium carpini*. Both species occur throughout Europe, although not in the UK, with *Priobium carpini* favouring the warmer, southern part of the range. *Priobium carpini* has also been found in the USA (Ford, 1984) in a large stack of rotten timber stored within the hold of the *USS Constellation* in Baltimore harbour, and in woodland near Baltimore airport.

Asia has different species within the familiar genera, but there is still a substantial overlap. Thus Hsu and Wu (1989) found that *Ptilinus fuscus* was an important pest of poplar timber in Qinghai Province, China, whilst Espanol (1969) records the same species as an occasional pest in Spain.

This brief review is not intended to be exhaustive, but the intention is to show that a worldwide problem is caused by a few beetle species, and that these beetles are easily carried from country to country in infested materials.

TIMBER DURABILITY AND INSECTICIDE TREATMENTS

In the last section we have discussed how only a few beetle species are a significant problem in buildings, although there are many more anobiids with similar habits in the forests of the world. Many of these are occasionally found in buildings, and they provide a considerable resource of potential wood-destroying organisms if circumstances change. An example of this is found in buildings in Lhasa (Tibet) which were apparently traditionally constructed from cypress wood and other durable timbers. These timbers are no longer available, and poplar is now used for repairs. The consequence has been severe damage by a previously almost unknown beetle *Nepalanobium quadricostatum* (Ridout, unpublished data).

A change in timber quality may also have important consequences. This is especially true of softwoods where rapid plantation growing produces different characteristics to those of traditional wild-grown wood, particularly a far greater percentage of vulnerable sapwood.

The susceptibility of the new wood to beetle attack needs to be evaluated for two reasons:

- repairs to historical buildings should be made with the most durable material available if possible (of the same species)
- today's modern buildings will be tomorrow's historical structures

Loss of quality in modern softwoods is not generally perceived as a problem because the wood will readily accept preservatives. It seems preferable for long-term conservation to select wood for natural durability, rather than to rely on impregnation with chemicals of doubtful permanence.

Table 1 Timber moisture content requirements (MC) of woodboring anobiid beetles

Name	MC (%)	Author
Hadrobregmus carinatus	20.5★	Simeone (1961)
Ptilinus ruficornis	16.9★	Simeone (1961)
Hemicoelus gibbicollis	13–19★★	Suome and Akre (1992)
Xyletinus peltatus	11.6 ± 0.7★★★	Williams (1983)
Anobium punctatum	12–13.5★★★	Becker (1942)
Oligomerus ptilinoides	11–16★★	Cymorek (1979)

★ mean moisture content of infested wood
★★ moisture range
★★★ minimum MC

The treatment of active beetle infestation in readily accessible thin-section wood is easily accomplished by spray treatment with insecticides. If, however, there are applied finishes, such as gold leaf or lacquer, which resist absorption, then the problems become severe. Small items might be placed in a freezer, but large and immovable items, like church screens, might be almost impossible to treat. This type of problem will occur in historical buildings throughout the world. The only solution currently available is to fumigate the building with gas. This is always immensely expensive, impractical in many countries, and does not protect against re-infestation. The latter factor is important in cities like Istanbul, where the humid environment naturally favours the insects and both *Nicobium castaneum* and *Oligomerus ptilinoides* cause severe damage.

The deathwatch beetle (*Xestobium rufovillosum*) presents a similar problem because the oak it infests resists chemical penetration and the infested timbers may be largely inaccessible within the structure. Heat treatments for buildings are available, but are unlikely to be a practical method except in exceptional situations, if only because a small church costs about £100,000 to treat. Many historical towns in the UK have signs of deathwatch beetle infestation in every building that predates 1700, and damage is plentiful in ancient churches throughout the land.

It is clear that an entirely different approach to beetle prevention and control must be formulated if we are to protect historical buildings, and those that will become the historical structures of the future. Recent research by Ridout Associates and the Jodrell Laboratory (Royal Botanic Gardens, Kew) has focused on two of the most important beetles species (*Anobium punctatum* and *Xestobium rufovillosum*), but our results will be applicable to other species around the world. We have accepted that timber and finishes may present a significant obstacle to bringing insecticides into contact with beetle larvae, and we have therefore concentrated on attracting and trapping the beetles as part of an integrated pest management (IPM) approach. This will deplete the population's reproductive potential, deflect foraging insects before they colonize, and monitor for current activity.

AN INTEGRATED APPROACH TO BEETLE CONTROL

Vast amounts of chemicals are sprayed on to the timbers in buildings every year where infestation is extinct, or as precautionary treatment where the actual chemistry of the timber makes attack impossible. An integrated approach to beetle control must therefore commence by assessing the risk of progressive damage.

Assessing current activity

Active infestation by many woodboring anobiids, such as *Anobium,* is indicated by fresh bore dust, which trickles from emergence holes. Detection by this method is useful, but not infallible because vibration, or even mites (Hallas, 1996), can dislodge the dust.

An active infestation of deathwatch beetle can be far harder to detect. If they are attacking sapwood then dust is usually present, but if the beetles are living in decay at the back of a plate or in the centre of a beam then they may emerge through old flight holes. These may be some distance from their larval feeding site, and bore dust will probably be absent. Pasting tissue paper or stapling card over groups of holes has proved useful for determining current activity, but the paper or card must be firmly attached. Old anobiid emergence holes in furniture or panelling can be clogged with wax polish immediately prior to the emergence season. In either case, the emerging beetles will make a hole in the covering material, wax or wood, and thus reveal their presence.

Precautionary treatment

Most of the timbers within a normal roof in the UK dated prior to about 1900 are largely immune from beetle attack. Precautionary treatments, except in exceptional situations, are worthless. Beetles restrict their attack to sapwood, unless the heartwood is either of a perishable species, or the wood chemistry has been modified by fungus. Ridout Associates are currently engaged in investigating the chemistry of infested pine timbers, and in every example we have analysed, the furniture beetle damaged portion has a far lower phenolic extractive content than the remainder of the timber, and thus may be characterized as residual sapwood. Damage in residual sapwood within a historical roof can usually be recognized, because the emergence holes will be restricted to one or two sides of the timber depending on how the wood was cut. This type of damage rarely requires treatment.

Modern forestry practice does not allow pine trees to grow wide stems before they are felled. The volume of sapwood within any part of the tree tends to remain constant, so that the thickness of the sapwood band in a log will depend on its diameter. Thinner logs have thicker sapwood, and are therefore a better food source for beetles. A modern roof, constructed from thin-section sappy softwood, may need extensive treatment if infested. An 18th or 19th century pine roof, constructed from wild-grown imported softwood, rarely requires insecticide treatment, if at all. Beetle larvae may eat wood, and a roof may be constructed out of wood, but it does not follow that the larvae can destroy the roof.

Insect behaviour and chemical treatments

Surface applied insecticides may kill insects in sapwood, but spray treatments have little, if any, impact on deathwatch beetle populations deep within heartwood. They were thought to be appropriate because laboratory investigations showed that eggs were laid on the surface of timbers, and that the larvae wandered extensively before burrowing. All stages of the insect were accessible to the insecticide. Hickin (1975) perceived an advantage to the insects in this behaviour because the eggs did not need to be laid at the feeding site.

The insect's behaviour in buildings is, however, rather different. The beetles emerge through old flight holes or shakes, and then after mating, the female beetle either re-enters the timber to lay her eggs or disperses to a new feeding site. Only a small percentage of the beetles bite their own emergence holes, and there is little contact between the insect and the surface-applied chemicals. Therefore, all insecticide treatments should be targeted and applied so that they penetrate deep into the timber.

Natural predators

Many books that discuss woodboring activities list clerid beetles, hymenoptera or mites as natural predators or parasitoides. All seem to ignore the house spider. Field observations demonstrate that spiders destroy large numbers of deathwatch beetles. The two genera commonly implicated are *Tegenaria* and *Pholcus*, but this probably reflects the frequency with which these spiders are found in buildings, rather than any particular preference for the beetles as a food source. Spiders are mostly opportunist feeders, and its seems likely that any spider of a suitable size may attack the beetles.

The two modes of attack normally encountered are entirely dissimilar. Female *Tegenaria* spp., when adult, form platforms of silk and live in a tunnel at one end or corner. Males and subadult females tend to be more nomadic. *Tegenaria* therefore trap beetles in webs or pounce upon them. The spider's chelicerae seem generally to pierce the beetle's elytra, and the insect is then dismembered. Fragments of beetles destroyed by *Tegenaria* are common in

roof spaces inhabited by the beetles. Frequently these are assembled into a neat pile beside the web.

Pholcus has a different mode of attack. The spiders weave a loose mesh of threads in which insects become partially ensnared, and the process is completed by the spider vibrating the web. *Pholcus* victims remain whole and are cocooned in silk for future consumption. These spiders and their prey are frequently suspended from ceilings or the undersides of furniture, for example church pews.

Food preference and spider size

Two questions seem to be relevant.

- Is the food choice of spiders governed by availability or preference?
- What size do spiders need to be before they can attack the beetles?

Twelve wild-caught *Tegenaria domestica* of various sizes were allowed to become established in small plastic boxes and fed on fruit flies. These spiders were then starved for a week and each was offered one active deathwatch beetle. Two days later a second beetle was offered.

The results are presented in Table 2. Size was difficult to establish accurately without disturbing the spiders, but was measured with a ruler as total abdomen plus cephalothorax length when the spiders were stationary and in a convenient position within their containers.

Spider 11 was offered two more beetles after a further week had elapsed, in case some beetles were distasteful, but the spider remained disinterested.

Beetles are largely ignored until the spider is about half-grown and able to cope with them. Of the 15 beetles offered to spiders of 7mm or over, ten were eaten, one was ignored and two repelled the spiders. The latter may have

Table 2 Reactions of *Tegenaria domestica* to active deathwatch beetles

Spider No.	Size (mm)	Beetle eaten	Spider repelled	Beetle ignored
1	3			● ■
2	4			● ■
3	4			● ■
4	6			● ■
5	6		●	■
6	7	● ■		
7	8	● ■		
8	8	● ■		
9	8	● ■		
10	8	●		■
11★	14		●	
12	14	■	●	

● first trial

■ second trial

★ no trace of second beetle, may have escaped

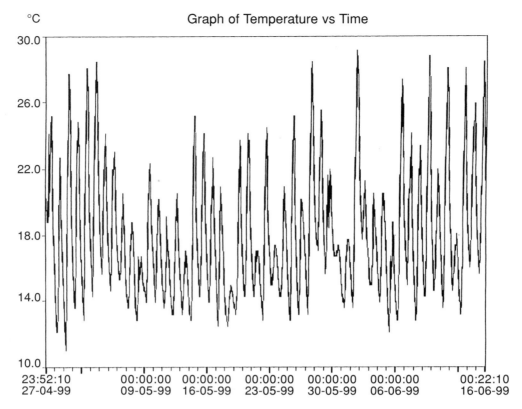

Figure 1 Air temperature in the nave roof at St. Michael's Church, Brent Knoll in Somerset

been a reaction to a novel prey. There may be an element of individual preference involved, because spider 11, having refused beetles, fed voraciously on flies.

Wise (1993) stated:

A predator has the potential to regulate a prey population only if the predator responds to increases in prey density by inflicting greater percentage mortality.

This type of response has rarely been convincingly demonstrated for spiders, and in the present case, feeding two beetles to spider 10 resulted in one beetle being eaten whilst the second was ignored. Feeding responses at higher prey densities are still to be studied, but the spiders tested were not killing more beetles than they could eat, although they would presumably be feeding at their maximum level.

Spiders are unlikely to have much of a controlling effect on a deathwatch beetle population that is flourishing in an ideal environment, but conditions are not always ideal. The deathwatch beetle larvae will live at timber moisture contents within the range tolerated by *Anobium* and other drywood anobiids, but they flourish in damper conditions. Drier conditions extend the larvae growth which produces smaller adults which are less reproductively viable and successful (Ridout, 2000). Beetle populations seem to fragment into small colonies where conditions are most favourable, and natural predation may become significant in this situation. Incautious general insecticide treatments may deplete the foraging predators whilst having little effect on the pest insects.

Trap depletion

Field observations strongly suggest that beetle populations will decline and eventually become extinct if timbers are kept dry in a well-maintained building. Adult beetle depletion should assist the process. Natural predators are useful, but beetle trapping can be more effective, particularly with deathwatch beetles.

The general opinion throughout most of the 20th century was that the deathwatch beetle rarely, if ever, flew. Baker (1964) found that a heat source would make the beetles fly from a surface warmed to 30–40°C at an air temperature of 22°C. The English Heritage Woodcare research found that beetles flew vigorously at an air temperature of about 17°C.

The following case study is informative. In 1996, 40 dead beetles were picked up from the central walkway along the Nave roof of St. Michael's Church, Brent Knoll in Somerset. These clearly indicated deathwatch beetle infestation, but not the extent of the problem because most of the vaulted ceiling was inaccessible for searching and the beetles may have been present for several seasons. The roof was enclosed and covered with lead, so that it was reasonable to suppose that the interior would reach temperatures that were significant to promote flight (Figure 1). We therefore installed a trap equipped with two 40 W UV lamps during the emergence season.

The trap has been used now for four consecutive years, and each beetle was sexed by dissection at the end of the emergence period. The results are shown in Table 3.

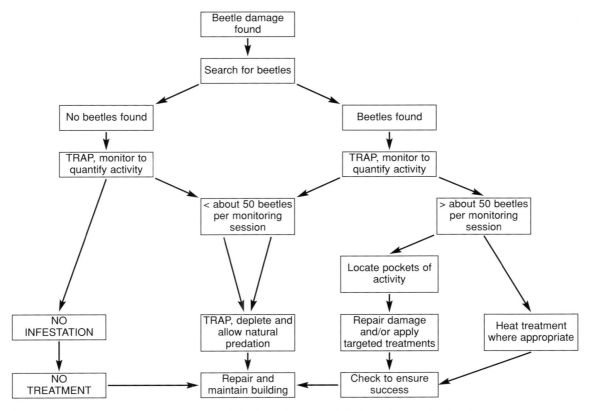

Figure 2 The integrated pest management of deathwatch beetle (*Xestobium rufovillosum*) infestations

Table 3 Annual catch of deathwatch beetles from St. Michael's Church, Brent Knoll in Somerset

	Male	Female	Total
1997	1	168	169
1998	0	230	230
1999	22	180	202
2000	4	238	242
Dead on central walkway in 1997	19	15	34

The central walkway had been swept clean of beetles prior to the 1997 season, so that the 34 beetles found were all freshly emerged. The sex distribution of the 34 was about even, but the trap catch each year is predominately female beetles. Most are full of eggs. Evidence from this and other finds suggest that females are attracted to UV light or are more free-flying than males.

CONCLUSIONS

The woodboring anobiid beetles may be loosely grouped into those that require damp conditions and fungus, e.g. *Priobium carpini* and *Hadrobregmus pertinax*, and those that tolerate dryer timbers, e.g. *Anobium punctatum* and *Oligomerus ptilinoides*. The damp-loving anobiids are generally easy to control by drying, but those that tolerate dryer conditions can be far more difficult to deal with. Surface applications of insecticides can be effective, but is not always practical or desirable, and in many cases an integrated approach to pest management will be

appropriate. A scheme devised for the control of the deathwatch beetle is shown in Figure 2. The underlying principles are:

- a building should be dry and well maintained
- infested material and debris, which might provide a reservoir for infestation, should be removed as far as is practical
- chemical treatments should be applied only if there is significant current activity, and these treatments should be targeted
- traps should be considered for the evaluation of suspected beetle activity, and the depletion of beetle population

REFERENCES

Baker J, 'Flight behaviour in some anobiid beetles', in *Proceedings of 12th International Congress in Entomology*, 1964, London, 319–320.

Becker G, 'Okologische und physiologische untersuchungen uber die holzzerstorenden laervan von *Anobium punctatum*', in *Okol Tiere*, 1942, **39**, 98–151.

Cymorek S, 'Uber den Nagekäfer *Oligomerus ptilinoides* (Wollaston), Col., Anobiidae: Verbreitung und Einschleppung, Bestimmung, Lebensbeziehungen und Befallsmerkmale mit Vergleichen zu *Nicobium*', in *Material und Organismen*, 1979, **14**(2), 93–104.

Espanol F, 'Anobiides nuisbles en Espagne au bois ouvre (Coleoptera)', in *Memorie della Societa Entomologica Italiana*, 1969, **48**(1B), 25–28.

Ford E, 'Priobium carpini (Herbst) an Old World anobiid now established in Maryland (Coleoptera)', in Coleopterists Bulletin, 1984, **38**(3), 249–250.

Hallas T, 'Dust from woodworm holes may also be a result of house mite activity', in International Pest Control, 1996, **38**(5), 157.

Halperin J, 'Occurrence of Oligomerus ptilinoides in Israel', in Phytoparasitica, 1992, **20**(1), 51–52.

Hickin N, The Insect Factor in Wood Decay (Third ed), 1975, Associated Business Programmes, London.

Hosking G, 'Anobium punctatum (De Geer) – Houseborer', Forest and timber insects in New Zealand, 1978, **32**, 7.

Hsu C, Wu H, 'Bionomics and control of the anobiid beetle Ptilinus fuscus', in Quinhai Acta Entomologica Sinica, 1989, **32**(2), 200–206.

Mauldin J, Barry S, Houghton D, 'A review of the most important termites and wood destroying beetles in the United States', in 6th International Biodeteriorations Symposium, August 1984, 1986, Washington DC, USA, 667–671.

Muirhead D, 'A beetle control problem in timbers of the Old South Meeting House', in Journal of Economic Entolomogy, 1941, **34**(3), 381–383.

Nour H, 'Anobiidae attacking furniture', in Bulletin Entomolugique d'Egypts, 1962, **46**, 1–383.

Peters B, Fitzgerald C, 'Anobiid pests of timber in Queensland: a literature review', in Australian Forestry, 1996, **59**(3), 130–135.

Ridout B, Timber Decay in Buildings, 2000, Spon, London.

Simeone J, 'Observations on Hadrobregmus carinatus (Say) and other wood-feeding Anobiidae (Coleoptera) in the North Eastern United States', in Dissertation Abstracts, 1961, **21**(7), 2060.

Suomi D, Akre R, 'Distribution of economically important, wood infesting anobiid beetles in the Pacific north-west', in Journal of the Entomological Society of British Columbia, 1992, **89**, 57–62.

Suomi D, Akre R, 'Biological studies of Hemicoelus gibbicollis (Leconte) (Coleoptera: Anobiidae), A serious structural pest along the Pacific cost: Larval and pupal stages', in Pan-Pacific Entomologist, 1993, **69**(3), 221–235.

USDA, 'Fascicle of the catalogue of the Coleoptera of America north of Mexico, 1982', Agricultural Research Service Handbook, United States Department of Agriculture, 1982, 529–570.

Veronese M, 'Difendiamo le strutture in legno della casa di campagna dagli insetti parassiti', in Vita in Compagna, 1995, **13**(4), 55–57.

Williams L H, 'Wood moisture levels affect Xyletinus peltatus infestations', in Environmental Entomology, 1983, **12**(1), 135–140.

Wise D, Spiders in Ecological Webs, 1993, Cambridge Studies in Ecology, Cambridge University Press, Cambridge.

BIOGRAPHY

Dr Brian Ridout read applied biology at Cambridge University where he obtained an MA in entomology and mycology. This was followed by a period of research at Birkbeck College, University of London, which produced a PhD in entomology. He has now been elected an Honorary Research Fellow.

In 1987, he established his own company and was commissioned as English Heritage's timber decay consultant. In 1996 he was commissioned by English Heritage and Historic Scotland to write a text book which would underpin the Heritage approach to the control of timber decay organisms in buildings. His book, Timber Decay in Buildings: The Conservation Approach to Treatment was published in 2000.

CHAPTER FIVE

THE DEVELOPMENT OF AN INTEGRATED PEST MANAGEMENT POLICY FOR THE NATIONAL MUSEUMS OF SCOTLAND

Isobel Griffin

National Trust for Scotland, South Regional Office, Northgate House, 32 Northgate, Peebles EH45 8RS, Scotland
Tel/Fax: +44 1721 724 700   e-mail: igriffin@nts.org.uk

ABSTRACT

The National Museums of Scotland (NMS) has a large and diverse collection, spread across a number of separate museums and stores. Some areas of collection care have progressed very well over the last 20 years or so, but until recently pest management had only been partially addressed. A new post of preventive conservator was created in September 1999, with a job description including the development of pest management within NMS. The first task was the creation of an integrated pest management (IPM) policy, achieved through consultation with several other large museums and an independent pest consultant. Implementation of the policy began in spring 2000, and although some changes were easy to introduce, other issues posed problems which are still being resolved. However, in terms of raising staff awareness the policy has been a great success.

KEYWORDS

Pests, management, policy, housekeeping, trapping, communication, treatment

INTRODUCTION

The National Museums of Scotland (NMS) has a large and diverse collection spread across a number of separate museums and stores. While some areas of collection care, such as environmental monitoring and materials testing, were well developed, pest management had only been partially addressed. The Geology and Zoology department had a strict quarantine procedure for incoming specimens, whereas objects entered other departments with little or no checking. Insect trapping was only undertaken in areas of particular concern and although a chest freezer had been purchased, other potential treatment methods had received insufficient consideration following the discontinuation of treatments in the museum's low-pressure methyl bromide chamber.

Despite these limitations, NMS has been fortunate in avoiding major pest infestations. However, in August 1998, an infestation of *Trogoderma angustum* (cabinet beetle) was found in closed glazed casing used to store mounted birds in the Royal Museum, Chambers Street, Edinburgh (Shaw, 1999). The gallery had not been open to the public for eight years and since that time the glass of the cases had been papered over internally and the contents left undisturbed (Figure 1). Hundreds of insects were found, but fortunately, the outward signs of damage to the bird specimens were fairly slight. A search of the surrounding stores and galleries revealed one other infestation in a primates case in the gallery below, which is connected to the floor above by an open well.

The insects had probably entered the building through a pane of glass above the bird cases, which was broken and not repaired for several months, about six years prior to the infestation being discovered. They were eradicated through a thorough clean-up operation and the localized use of pesticides. The cases have since been monitored closely.

This incident and the general awareness of risk elsewhere to the collections highlighted the need for more coordinated pest management. When a new post of preventive conservator was created in September 1999, this was included as a significant part of the job description. It was decided that an integrated pest management (IPM) policy was required, to clearly state the NMS position and detail appropriate procedures. As IPM coordinator, the preventive conservator was responsible for writing the policy and facilitating its implementation.

Although no specific budget was created for IPM, the preventive conservator received a start-up budget, which was used in part to purchase crucial equipment and materials, such as insect traps. The relatively small sums of money needed to replace traps are now included in the preventive conservation budget, but larger sums of money, for example, to purchase equipment for undertaking pest treatments, have to be bid for separately.

Figure 1 The run of cases at the Royal Museum in Edinburgh, used for the storage of birds in which the infestation of *Trogoderma angustum* was found. © NMS

BACKGROUND RESEARCH

Given that several other museums were already known to have developed their own IPM policies, it was considered desirable to benefit from their experience. The IPM co-ordinator therefore made visits to several museums, speaking to conservators and curators responsible for a range of different collections. The aims were to establish how IPM operates inside other museums and also to compare the appropriateness and success of the various pest treatments in use.

All of the museums visited had an IPM policy and a committee to implement the policy. The Natural History Museum in London, in particular, had a very well organized system of committees and sub-committees. The experiences of the conservators at the National Museums and Galleries on Merseyside, in trying to coordinate pest management throughout a group of museums, were very relevant to NMS. Not surprisingly, the need for involvement at every level in the museum was frequently stressed.

Freezing was the preferred method of treatment both at the Natural History Museum and at the National Museums and Galleries on Merseyside. However, the Natural History Museum also uses pesticides at a low level to reduce the risk of infestations, and the National Museums and Galleries on Merseyside stressed that not all objects can be frozen. The Victoria and Albert Museum freezes many of its textiles, but is also developing anoxic treatments using nitrogen for treating furniture. The Horniman was the only museum I consulted with which still uses the methyl bromide treatment on a large scale, and it was therefore interesting to hear how they are planning to replace it.

IPM POLICY

Following the visits to other museums and discussion within NMS about what was feasible and workable, an IPM policy was drafted, with the help of an external consultant, David Pinniger. The draft was circulated to the department heads and the Head of Collections, and their comments were used to produce the final version. The paper and procedures were then approved by the NMS Management Board and it became official museum policy.

The IPM policy consists of an introduction explaining the purpose of the policy and a list of key policy statements, followed by more detailed guidelines for implementation of the policy. The guidelines cover the structure of the IPM group, staff training and awareness, housekeeping practices, pest monitoring, incoming objects and treatment of infested objects. There are appendices detailing the responsibilities of all staff, and the procedures followed by the IPM coordinator for inspecting and treating incoming objects, and dealing with findings of insects, birds or rodents.

The IPM policy is a key element of the NMS Collections Management policy, which also incorporates documents such as environmental policy and the storage and handling policy.

The IPM group

The IPM group consists of the IPM coordinator and deputy coordinator (preventive conservator and furniture conservator), the Head of Geology and Zoology, the Head of Conservation and Analytical Research and the Head of Estates. There are obviously other staff involved in IPM, such as the Health and Safety Officer, the Cleaning Operations Manager, the Head of Visitor Services and the representative for Events and Hospitality. However, it was decided that these staff would be kept informed of the group's activities through circulation of the minutes, and would be invited to attend individual meetings as appropriate.

The meetings have proved useful for sharing information and ideas from different departments. However, although the initial intention was to hold quarterly meetings, in practice, there has only been sufficient material for discussion to warrant meetings twice a year.

Training and awareness

To achieve successful implementation of the IPM, it was necessary both to raise general levels of awareness about pests and to provide specific training for staff. Following acceptance of the IPM policy by the management board, an initial presentation was given to the Department of Conservation and Analytical Research. This meant that the pool of staff available to promote the policy and answer queries was larger than just the IPM coordinator and deputy coordinator. The policy was then introduced museum-wide at a staff presentation, for which attendance was compulsory.

Less formal presentations included a 'curator's choice' talk, entitled 'The enemy within: woodworm, pests and bugs in the museum', and a light-hearted talk given by David Pinniger during his initial consultation visit. The attendance of key members of staff such as the Head of Estates and the representative for Events and Hospitality was achieved by issuing them with personal invitations.

Other communication strategies have been employed such as group e-mails, preparing staff for 'flying time' in the spring and announcing exciting insect finds within NMS. Besides containing useful information, the e-mails have alerted staff to the existence of the IPM coordinator. As a result, there has been a surge both in general communications regarding pests, such as a recent spate of

mouse sightings, and in the number of insect specimens arriving for identification. In addition, NMS has a procedure for circulating important information in museum-wide general notices, and one of these was produced to explain the procedure for requesting freezing treatments for high-risk objects. Following circulation of the notice, there has been an increase in the number of requests for freezing, but staff have been encouraged to wrap their own objects and make transport arrangements.

One of the training aims in the policy was that the staff most closely involved with pest management would attend training courses. We have run a two-day IPM course with David Pinniger at NMS in 1999, and again in 2001, where it has been popular with both conservators and curators. Besides facilitating external courses, we have built up a collection of teaching aids such as slides, insect specimens and unregistered objects, to facilitate the provision of in-house training.

Although we appear to be getting the message across to staff at a certain level, we have probably had less success with large groups such as the cleaners and Visitor Services staff. These staff do not have access to e-mail, and are generally not available to attend talks because they work shifts. It has proved best to target them individually, for example, during walkabouts to point out the insect traps and explain their purpose (Figure 2). However, in such a large institution with a high turnover of staff, this training needs to be repeated at regular intervals, and we do not always have the necessary resources.

Figure 2 A conservator shows the contents of a sticky trap to a staff member from Visitor Services. © NMS

Housekeeping practices

Increased awareness about pests has generated some improvements in housekeeping practices. The previous practice of schools being allowed to eat their packed lunches in the corner of the ethnography gallery has been stopped and a dedicated eating area has been created, which is further from museum objects, and more accessible for cleaners. Another example is the proposal put forward by the textile and paper conservators to install fine-mesh fly screens on their laboratory windows.

Areas still requiring more work are the revision of existing cleaning routines and the implementation of additional specialist cleaning. Everyday cleaning is undertaken by cleaners, who typically sweep and wet mop the floors. Vacuum cleaners are only used in certain areas, and many of the conservation and curatorial staff are unhappy with the lack of vacuuming. This is mainly because they are concerned that dust and any insect pests present are not being effectively removed, and also because the wet mops may mark objects on open display. Although the Natural History department has managed to implement some improvements to the cleaning regimes in their stores, it has generally been quite difficult to introduce changes. This is an area that we have targeted to develop further.

Specialist cleaning is also an issue. Objects which are on display are generally cleaned by curators although the thoroughness of this varies from one department to another. However, cleaning and checking of objects in storage, and deep cleaning of the stores themselves can be very infrequent, and it is these collections which are probably most at risk.

A pilot specialist cleaning and pest-monitoring programme was undertaken by the preventive conservator and the furniture conservator at Leith Customs House, which is used mainly for furniture and textile storage. The work included checking and cleaning objects, deep cleaning stores and implementing simple improvements to improve the storage conditions. For example, new dustsheets were purchased to protect all objects, and old air vents and fireplaces were covered with plastic. In addition to the obvious benefits, it was hoped that the programme would help identify the resources needed to undertake similar work elsewhere and also that it would encourage other staff to become involved.

It took approximately 12 staff-days to work through the 18 ground floor rooms. The material costs were borne by the relevant curatorial departments, who were very supportive of the work, and the results have generated a lot of positive feedback. However, more involvement from other conservators and curators is still required. Deep cleaning and object checking needs to be repeated annually at least, and while the preventive conservator and furniture conservator may be able to maintain standards at Customs House, they do not have the resources to take on all the other stores.

Faced with such an enormous task, the preventive conservator has focused on the areas where help has been positively requested. Thus, in addition to the annual deep clean of Customs House, an annual deep clean of the stores at the National War Museum of Scotland has begun in collaboration with an assistant curator from the Social and Technological History department. Also annual cleaning of some of the cellar stores is being undertaken with assistant curators from the History and Applied Arts department. It is anticipated that the deep cleaning programme will gradually be extended, and that the preventive conservator will be able to adopt a more supervisory role once routines have been established.

Pest monitoring

Pest monitoring is an area in which we have made good progress. The aims of the policy were that all staff should be vigilant for pests and aware of reporting procedures, conservators and curators should check objects regularly and instigate special surveys as appropriate and insect trapping should be undertaken at a low level throughout NMS. There has certainly been an increase in the reporting of pests. Insect pests are sent to the entomologist responsible for pest identification, and sightings of mice and other vermin are reported to the Estates department. A database is kept, thus allowing 'hot spots' for pest activity to be highlighted.

The trapping is also well underway. We now have traps in all of our museums and stores with a total of around 150 sticky traps. We also use around 20 moth pheromone traps and furniture beetle pheromone traps which, for financial reasons, are only employed in high-risk areas. We only use traps in the galleries where we can find good hiding places for them, because previous experience has shown that portable items visible to the public are highly susceptible to disturbance or theft. Even so, traps still go missing, and it is likely that museum employees are mainly responsible for this.

The traps are checked on a quarterly basis by the IPM coordinator and the deputy IPM coordinator and any traps containing potential pests are removed and sent to the entomologist responsible for pest identification (Figure 3). The entomologist reports back to the IPM coordinator, and the results are entered onto a database. With so many traps, the inspection procedure is quite time-consuming, but we decided not to involve more people, due to concerns that this might generate disparity in the results.

So far the traps have not revealed any great causes for concern. Although there have been isolated examples of insects such as clothes moths and spider beetles on the traps, they have only ever occurred singly, and are therefore more likely to be random interlopers from outside, rather than evidence of infestations within the museum. We have not found any more *Trogoderma angustum* despite allocating several traps to the area of previous infestation, and we are therefore optimistic that it has been successfully eradicated

Figure 3 The preventive conservator and the entomologist responsible for pest identification examine an object showing signs of insect damage. © NMS

from NMS. However, *Trogoderma* are still very common at the Royal Botanical Gardens in Edinburgh, and have been found in the home of one of our natural history curators, indicating that we have no room for complacency.

Although trapping is time-consuming, it has been one of the easiest parts of the IPM policy to implement, and is a good way to bring IPM to the attention of staff, and to make them feel that positive action is being taken in the museums.

Quarantine and treatment

The IPM policy states that:

all incoming objects will be taken directly to a quarantine area upon arrival at NMS. They will be inspected and treated against pests before they are allowed to proceed into the museum.

We are working towards having a dedicated building among our stores at Granton with packing and quarantine areas, and a walk-in freezer. The vast majority of incoming objects will then be taken to Granton, where they will be inspected by a conservator and treated appropriately.

All suitable objects will be frozen, and objects not suitable for freezing will receive other in-house treatments, such as spraying with pesticide, or anoxic treatments with Ageless®. Objects which cannot be treated in-house will either have their treatment contracted out, for example, to Thermo Lignum® or Rentokil, or will be kept in sealed bags in the quarantine area until their safety can be guaranteed.

Practically, it is not feasible to insist that all objects enter the museum via Granton, as there will always be 'one-off' objects that are required almost immediately. We are therefore planning to establish an additional smaller quarantine area, containing a chest freezer, in the museum complex at Chambers Street.

Our proposals for the quarantine and treatment of incoming objects are being introduced using a phased approach, as we gradually build up our resources for this aspect of IPM. For example, we need to increase the human resources available. At present the IPM coordinator, deputy coordinator and the conservator responsible for freezing objects at NMS all have many other responsibilities, and if we were to insist upon a strict quarantine procedure the workload would be too great for the Conservation department. One possibility is to ask for some input from curators, but this in turn throws up issues of providing training and achieving a consistent approach.

Secondly, we need to accurately assess the demand that a strict quarantine procedure would place on our current freezer. Our chest freezer is definitely not big enough to deal with the individual sizes of some objects, and it may also be too small to accommodate the quantity of objects suitable for freezing. The possibility of acquiring a walk-in freezer for NMS has been discussed, and given that such a facility is currently unavailable in Scotland, it would undoubtedly prove a valuable resource for other Scottish institutions. However, before we look for the funding (approximately £15,000), we need to verify whether a walk-in freezer is really required. This we are doing by keeping records of all incoming objects which are suitable for freezing.

The third issue is the funding for external treatments. The financial implications of arranging for all objects that are unsuitable for in-house treatments to be dealt with by, for example, Thermo Lignum® or Rentokil, are significant, and the cost benefits of this approach must be questioned. Given that we have historically encountered very few problems with contaminated objects entering NMS, and therefore the risk of this happening must be considered low, do we really want to spend our precious resources on costly external treatments?

A further problem is the acquisition and maintenance of sufficient 'empty' space at Granton and at the Chambers Street complex. There is a great demand for accommodation within NMS, and IPM must be prioritized at the highest level if treatment areas are to be made available.

Thus, given that we are still developing our resources, our current policy is to treat only high-risk objects. The largest group of objects requiring treatment so far has been the agricultural collection which, until recently, was stored in very poor conditions at Port Edgar in South Queensferry. Many of the wooden objects showed evidence of woodworm infestation, and although in some cases this was probably inactive, in others it was clearly alive. The collection was being moved to a new purpose-built museum at Kittochside near East Kilbride, and following discussions with David Pinniger it was decided to treat all wooden parts with Constrain® pesticide, and to freeze any leather or textiles. This required a structured programme over several months involving several conservators and curators.

The agricultural collection was considered a priority because it was moving to a new museum and hence was evidently a serious infestation threat. However, the inspection and treatment of other objects has been less consistent. For example, objects arriving at NMS and requiring conservation treatment are thoroughly inspected by conservators and given pest treatments as necessary. Similarly, staff from the Education department are acutely aware of the risk posed by objects returning from loan, and always arrange for treatments to be undertaken. However, other members of staff can be more blasé, resulting in some objects being moved straight into stores still in their packaging, and left without checking for many months or even years.

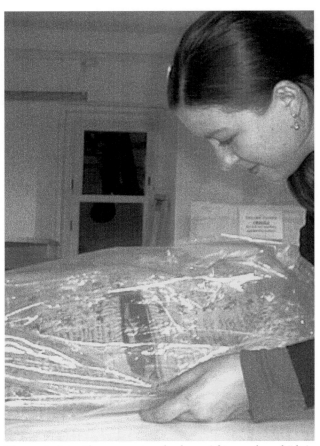

Figure 4 A textile conservator checks an African robe, which is undergoing an anoxic treatment with Ageless® oxygen scavenger. © NMS

More positively, we have collaborated with the Geology department to develop a good technique for anoxic treatments using Ageless®, which they utilize for the long-term storage of pyrite specimens (Figure 4). We can now confidently offer this as a service within NMS.

The current situation will only be fully resolved once we have sufficient resources to implement a museum-wide quarantine and treatment procedure. However, in the meantime it is important to educate staff. They should be aware of the need to check incoming objects, and they should know what facilities are available for pest treatments, and when it is appropriate to request these treatments.

Conclusions

Overall, we feel that we have made great progress with IPM at NMS. Our starting point in 1999 was that we had no policy, no individual or group of individuals responsible for IPM, sporadic trapping at only two of our sites and no coherent approach to the treatment of incoming objects, except in the Natural History department. There was a general feeling that housekeeping issues should be addressed, but this had only partially been achieved.

We now have an established policy with an IPM coordinator, deputy coordinator and a committee, and we have managed to raise awareness among staff and to begin a training programme. Communications between the cleaners and the other museum departments have improved. Housekeeping standards have also improved, with the introduction of deep cleaning regimes in several areas. We have a comprehensive trapping strategy across all of our sites, and we have established an initial quarantine area and a trial procedure for dealing with incoming objects, to be developed as we assess our needs and the additional resources required.

Our experiences have demonstrated that for IPM to be successful there must be involvement from staff right across the museum. In addition, IPM must be seen as a priority by senior management in order to make the necessary resources available. Good communication is vital for an effective policy, and this must be ongoing both to spread the message to new staff and to prevent complacency in others.

REFERENCES

Shaw M, '*Trogoderma angustum* (Solier, 1849) (Coleoptera: Dermestidae), a museum and herbarium pest new to Britain', in *Entomologist's Gazette*, 1999, **50**, 99–102.

MATERIALS AND EQUIPMENT

Constrain® (water-based pesticide with active ingredient permethrin)
  Historyonics
  17 Talbot Street
  Pontcanna
  Cardiff CF11 9BW
  Wales
  Tel: +44 29 2039 8943
  Fax: +44 29 2021 8672

Ageless® (oxygen scavenger containing powdered iron oxide)
  Conservation by Design Ltd
  Timecare Works
  5 Singer Way
  Woburn Road Ind. Estate
  Kempston
  Bedford MK42 7AW
  United Kingdom
  Tel: +44 1234 853 555
  Fax: +44 1234 852 334

ACKNOWLEDGEMENTS

The author would like to acknowledge David Pinniger for his advice, Sarah Gerrish for her work as deputy IPM co-ordinator, Mark Shaw and Andrew Whittington for undertaking insect identification and Jim Tate and Kathy Eremin for help with the text.

BIOGRAPHY

Isobel Griffin obtained an honours degree in Natural Sciences and the History of Art from Cambridge University, followed by a post-graduate diploma and a masters degree in the conservation of wall paintings from the Courtauld Institute of Art. She worked at the Courtauld Institute of Art as a research and teaching assistant and undertook a three-month internship in the Conservation Research department of the British Museum before joining the National Trust as an environmental monitoring and control intern. She was the preventive conservator at the National Museums of Scotland from 1999–2000, and has been a regional conservator for the National Trust for Scotland since January 2001.

CHAPTER SIX

MARAUDING GECKOS – A LOOK AT SUBTROPICAL PEST MANAGEMENT

Lydia Egunnike

CONSERVATION UNIT

State Library of Queensland, PO Box 3488, South Brisbane, Queensland 4101, Australia
Tel: +61 7 3840 7779   Fax: +61 7 3840 7943   e-mail: l.egunnike@slq.qld.gov.au

ABSTRACT

This paper discusses the complexities of implementing a functional Integrated Pest Management (IPM) programme at the State Library of Queensland, Brisbane, in Australia. Brisbane has a subtropical climate, which is highly conducive to pest and mould activity. Having been plagued by an exciting mix of mould and pest outbreaks in a short space of time, it was apparent that the library desperately required a comprehensive IPM programme to minimize future outbreaks. IPM programmes have been attempted previously but were not successful for a variety of political, structural and financial reasons. This paper will discuss past, present and future aspects of pest management at the State Library, with the view that past and present problems may be overcome to ensure a practical and sustainable pest management programme for the future.

KEYWORDS

Integrated pest management (IPM), sub-tropical, policy, training, geckos, cigarette beetle

IINTRODUCTION

Brisbane, located in the south-east corner of Queensland on the eastern coast of Australia, has a subtropical climate. The summers are hot and humid with diurnal temperatures usually between 20–30°C and humidity regularly over 60%. The highest rainfalls are normally in the summer months. Winters are dry and cool with diurnal temperatures from 5–20°C and humidity often as low 25%. The months from October to April are the most conducive to pest and mould activity, although the absence of a very cold winter means many pests are active throughout the year.

The State Library of Queensland is located on the south bank of the Brisbane River in Brisbane's central business district. It is part of the Queensland Cultural Centre, which is also home to the state's museum, art gallery and performing arts centre. Off-site library buildings include the Public Lending Division, which focuses on the regional lending programme, and the Dutton Park repository, which houses less current volumes from the State Reference collection and overflow collections from the Special Collections division. It is also the home of the State Library Microfilm Unit. The main library collections are housed at the principal site at South Bank.

THE BUILDINGS

The first line of protection against pest infestation is the building. As mentioned previously, the State Library of Queensland is located in three different buildings, but only two of these facilities, South Bank and Dutton Park, will be discussed in this paper. Not all building-related problems impacting on pest management programmes at Dutton Park and South Bank can be addressed here, so only the major issues will be discussed.

The main State Library building, located in South Bank on the Brisbane River (Figure 1), was opened in 1988. It is a four-storey reinforced cement building with the majority of external walls made of reinforced glass windows. Despite the building being only 13 years old, structural problems have already begun to emerge. A lack of appropriate storage space has been an ongoing problem with overflow collections being stored in unsuitable areas.

The use of full-length windows along the majority of external walls has created poor thermal insulation and light problems in a number of work and display areas. The glass doors along these walls are not properly sealed and allow access into the building for a variety of pests including ants and geckos. These gaps also allow heavy rain to get in.

It has long been acknowledged that a stable environment is an integral part of a successful pest management programme. Achieving the appropriate environmental set points in both the South Bank and Dutton Park buildings has proved problematic. When both buildings were designed, the temperature and relative

Figure 1 Aerial view of the State Library of Queensland (Photo courtesy of Gary Shepard)

humidity parameters were not as stringently set as today's requirements. Several of the existing air conditioning systems have difficulty maintaining the current rigid set points. The upgrading of systems in the South Bank building has been hindered by the fact that no space was provided in the plant rooms for the expansion and upgrading of plant equipment. This problem has been compounded by the fact that all external alterations to the building must be approved by the building architect. Plans to construct additional plant room space on the roof of the building have so far met with disapproval from the architect. This has made it impossible for our air-conditioning technicians to satisfactorily improve the current systems.

The inappropriate location of storage repositories has also placed great demand on the air-conditioning systems. The majority of repositories for historical collections are located along the poorly-insulated western wall, which, during the summer months, bears the full brunt of the afternoon sun. This causes temperature and humidity levels within the repository to fluctuate wildly. It is only a matter of time before these high temperature and humidity levels result in increased pest activity.

The problems at the South Bank building pale into insignificance when compared to the building problems faced by the Dutton Park store. The Dutton Park store was built in the late 1960s to house the Queensland State Archives; the State Library became a tenant in 1996 when the Archives moved to a new purpose-built facility. The building is directly managed by Q Build, a state government agency responsible for numerous government properties. There is no on-site manager, so many building and air-conditioning maintenance and repair problems are not dealt with, or are identified too late. From the beginning, it has been an ongoing battle to control pest and mould outbreaks at this facility. Large cracks in the building structure have provided easy access for a variety of pests including possums. Ongoing rodent activity has caused considerable damage to books in the State Reference collection.

The Dutton Park air-conditioning units have long been unsatisfactory and fluctuating conditions have encouraged a variety of pest outbreaks including carpet beetles, termites, cigarette beetles and cockroaches. Regular flooding occurs when poorly maintained drip trays overflow onto compactor below.

Recently the library received funding approval for a new library building, three times the size of the current site. It is crucial that mistakes in the design and construction that have had a detrimental impact on the pest management programme of our current facilities, will not be repeated. This time, conservation staff are heavily involved in the design and functional briefs, so there is hope.

THE COLLECTIONS

The State Library of Queensland collections provides a major information source for all Queenslanders. The types of material housed in the State Library of Queensland are typical of library and archive collections. Material is predominantly paper-based; either flat paper documents including maps, manuscripts and works of art on paper, or cloth, leather and card bound volumes. The library also houses a significant photographic collection containing a variety of photographic processes. Digital media is also becoming an integral part of many collections.

The collections are divided into two categories. The Reference Collection consists of over 5300 titles and is located on level two and part of level three of the South Bank facility. The collection comprises of non-original books, magazines, journals, microform, and digital media. Most material is housed at South Bank with older material stored at the Dutton Park repository.

The Special Collections Division contains a number of separate collections dedicated to specific subjects or media formats. It comprises the following units, located on levels three and four of the South Bank building. The Arts Unit contains the library's visual arts collections. The Unit has an extensive collection of rare books, manuscripts, and artworks. This material is housed in a temperature-controlled and humidity-controlled repository and reading room, both located within the unit.

The Audio-visual Unit, also on level three, is responsible for the division's general audio-visual collections. The collection contains duplicate video and audio-cassettes, motion picture film, and slides. Rare, historical material is the responsibility of the John Oxley Library.

The Family History Unit on level three manages a range of genealogical resources. The collection consists of books, reference works, microform, and CD-ROMS. The microform collection is housed in a temperature-controlled and humidity-controlled storage vault located within the unit.

Table 1 Table illustrating the most common types of pest activity, in the State Library of Queensland, January 1999–November 2000

Pest	Number of outbreaks	Risk to collections
Cockroach (*Periplaneta americana, Blattella germanica*)	13	high
Malaysian gecko (*Hemidactylus frenatus*)	11	low
Cigarette beetle (*Lasioderma serricorne*)	7	high
Silverfish (*Lepisma saccharina*)	8	high
Booklice (*Liposcelis corrodens*)	4	medium
Carpet beetle (varied) (*Anthrenus verbasci*)	3	high
Rodents	3	high

THE PESTS

Many of the pest problems faced by the library are universal, although we must also contend with some more unusual pests specific to our region. Table 1 illustrates the main pest problems currently facing the library. Mould outbreaks have not been included in this table. Reaction to gecko, cockroach, cigarette and carpet beetle activities have used the most time and resources.

Dealing with the Malaysian gecko (*Hemidactylus frenatus*) (Figure 2), has proved to be one of the more challenging aspects of the pest management programme. These pale pinky grey geckos are regular guests at the South Bank building and on a quiet evening, their distinctive 'chuck-chuck-chuck' can be heard throughout the stacks. The extent of the gecko population within South Bank has been difficult to assess, as conventional insect monitoring techniques such as trapping cannot be humanely used.

The Arts Unit on level three of the South Bank building is the epicentre of the library's gecko activity. The reason why the Arts Unit is a haven for geckos has not yet been discovered. The fact that the library is located directly next to a river may account for their presence in the building. They are known to frequent the wharf areas

Figure 2 A Malaysian gecko (or Asian house gecko) (Photo courtesy of Queensland Museum)

(Ryan, 1995) along the Brisbane River. However, it does not explain why they are most commonly seen in the Arts Unit, which is located on the third floor on the non-river side of the building. Further investigations are being carried out.

Although the geckos do not eat library materials, they can cause damage by the acidic white substance they excrete which can stain. This substance is uric acid, the metabolic by-product from the kidneys. Geckos are insectivores, therefore it is reasonable to assume that where there is a healthy gecko population there is also a healthy insect population. To successfully reduce the gecko presence within the Arts Unit, it was necessary to assess the extent and variety of insect activity in the area. Insect blunder traps were placed in a number of locations throughout the unit. The results of the trapping identified cockroaches (*Periplaneta americana, Blattella germanica*) as the main insect species in the area. One of the main entry routes for these cockroaches and geckos is through the gaps in the glass door, which adjoins an indoor garden area. Until these gaps are filled in, the insect population will continue to flourish. Eating in this area has been banned so insects are not being lured by food debris.

Cigarette beetles (*Lasioderma serricorne*) are a universal enemy to library collections. Large infestations of these beetles have caused significant damage to the library's legal deposit newspaper collection (Figure 3) located in our off-site store at Dutton Park. Unfortunately, despite freezing the affected collections and ongoing monitoring, this problem has not been solved and isolated outbreaks of beetle activity continue to be found. The unsuitable environmental conditions in the building and repository area where the newspaper collections are housed ensure the continuing nature of this problem. In the Dutton Park repository, cigarette beetle outbreaks are most likely to occur in the newspaper room and are most active during

Figure 3 A newspaper volume damaged by cigarette beetle larvae (Photo courtesy of the State Library of Queensland)

the summer months. The environmental conditions in this room are far from satisfactory, with only excess air-conditioning from adjacent temperature-controlled and humidity-controlled repositories reaching the area. The material most affected tends to be located in the compactor situated closest to the windows, which is exposed to the strong afternoon sun. The traps in this area are checked and replaced as necessary on a fortnightly basis.

Currently, most pest activity occurs at Dutton Park. However, at this early stage in the integrated pest management (IPM) programme it would be unsafe to adopt a complacent attitude to South Bank. Currently no widespread monitoring has been introduced in the building, so no data other than irregular entries into the pest logbook have been compiled. Dutton Park, on the other hand, is regularly inspected and two years worth of data has been collected and distinct trends have been identified.

PREVIOUS PEST MANAGEMENT APPROACHES

Pest management at the State Library of Queensland has previously been reactive in nature. The only regular preventative procedure was routine spraying of insecticides. This was not accompanied by any proactive approaches such as checking incoming materials. Pest management was not considered an integral aspect of preservation and so staff were only alerted to problems when epidemic levels were reached. Routine spraying of all library facilities continues.

Although the library itself did not have an operational pest management programme, in 1997, the Queensland Cultural Centre (QCC) management commissioned a pest audit (Gordh et al., 1998) of all cultural centre facilities, including the State Library of Queensland. This audit was carried out by staff of the Entomology department at the University of Queensland from June to November 1997. The only method of insect collection was blunder (glue) traps strategically located throughout the buildings.

Blunder traps were active for a six-month period, but were changed on a monthly basis. Samples were taken during June/July, July/August, August/September, September/October and October/November 1997. These periods were chosen to represent winter, spring and summer seasons. The contract did not allow for an autumn sample. The audit identified that insect abundance and diversity was subject to seasonal variations. The results inferred that the number of insects collected by traps in specified locations within different buildings represented seasonal abundance of outdoor insects (invasion of pests) or population abundance within buildings (resident pests).

The report states that pest control in QCC is exemplary. Within the State Library of Queensland building, the main problem areas identified were the food preparation areas of the staff tearoom and the public café, both located on the first floor. The conservation lab is also located on this floor near the café. Improved standards of hygiene and increased applications of insecticides were recommended in the audit. Strangely, no mention of implementing an IPM programme was made; in fact, the recommended approach relied heavily upon reactive techniques. Whether this report contributed to a feeling of complacency within the library management is difficult to ascertain. It did appear that despite some failed attempts by the Conservation Unit, no comprehensive approach to pest management had been implemented.

IMPLEMENTATION OF IPM AS LIBRARY POLICY

One of the keys to a successful IPM programme is institution wide cooperation at all levels. The attempt to become so far-reaching has proved to be a complicated process. It has taken two very expensive outbreaks to gain the attention of upper management. An IPM plan was submitted to the State Librarian in January 1998. At the time of writing, response to this plan has only just been received. It is hoped that 2001 will see a more far-reaching approach to pest control being embraced by all.

One of the most frustrating aspects of endeavouring to have an IPM approach accepted as library policy has been the inability to gain funding for pest management within the library. Presently, the conservation and general preservation budgets have supported ongoing pest control. Large amounts of funding were granted for the large-scale treatment of two significant insect and mould outbreaks at Dutton Park, both the results of ineffective pest management. It seems shortsighted that no additional funding has been provided despite the growing demands necessary to implement effective widespread trapping programmes in all three library facilities. Funding is also required to ensure appropriate non-chemical pest treatment such as freezing and anoxic methods. Simple equipment, such as edge sealers and industrial sized freezers, are needed. Presently we have to use outside facilities for even small outbreaks.

A proactive approach to pest management is undoubtedly more cost-effective than continual large-scale outbreaks, which can have negative financial and political implications. The public does not take kindly to the loss of access to collections their taxes have paid for. The library has an ethical responsibility to ensure any treatment undertaken on an object will not destroy its aesthetic or physical integrity, cause any permanent deterioration or compromise its research potential. The library's statutory responsibility for the care of the collections can also be called into question if material is damaged whilst in store, as occurred with the newspaper collections.

STAFF PARTICIPATION AND TRAINIING

It is accepted that IPM policies can only be implemented through the development of physical and operational changes. Staff need to know how they can help to minimize pest problems within the library environment. Education and communication are critical to success.

Presently, only staff involved in the Dutton Park store have been given pest training. It is planned that the remaining library staff will receive training in the first half of 2001. This means providing training for approximately 200 staff. In previous training sessions, groups have been small, with staff ranging from four to eight. Training consisted of a discussion and slide presentation showing common pests and the damage they may cause. Staff were provided with basic pest identification and inspection techniques. Each team was provided with a pest kit containing supplies necessary to carry out the inspection procedure. The items included standard inspection forms, torches, tweezers, specimen bottles, and bags. Supervised on-site inspections were carried out.

Initially, some staff were resistant to the programme. It was felt that such duties were not part of their official job descriptions. Once training was completed and staff understood the importance of the process, their enthusiasm increased. The routine inspection programme at Dutton Park has now been operating for a year and has prevented any further large-scale outbreaks.

Further staff involvement and training is required at the South Bank facility. Currently, staff in the Special Collection Units, particularly the John Oxley Library and the Arts Unit, are more versed in pest management issues than staff in the State Reference Branch. As these units contain some of the state's most valuable historically and culturally significant collections, a proactive approach to collection care has been adopted. Preventive preservation issues such as environmental control and safe handling techniques are already familiar to staff and many have had basic training in this area. There has been little problem with gaining the involvement and support in the pest management programme from collection managers and the majority of their staff.

The State Reference Division has an understandably different approach to collection care. Their main function is to provide a diverse and up-to-date reference collection to the public. Much of their collections are replaceable and do not warrant the particular care taken in the Special Collection units. It has been very challenging to introduce the concepts of preventive care to this division. At the time of writing, there is still a long way to go.

One of the biggest hurdles to conquer involving staff cooperation is the consumption of food in inappropriate areas. Currently, food consumption has not been successfully limited to the tearoom and café areas. It has ceased in the conservation area and in the Arts Unit, but remains a problem throughout the rest of the building. A major reason for this is the design of the building. All staff eating facilities are located on the ground floor. Staff on the upper levels often have insufficient time to go to the ground floor and will eat hurriedly at their desks between rostered duties. Their waste is placed in the open bins located under desks. These bins then sit until the morning when the cleaners empty them. This is a major cause for concern, particularly in areas where staff are located next to collection storage areas.

Indoor plants are another difficult issue to resolve, with staff unwillingly to relinquish their 'bit of green'. At this stage, a compromise has been made and as long as plants are healthy and show no signs of insect activity, they have been allowed to stay. Staff are relied upon to report any problems, but this does not always happen. Contract staff, responsible for watering the plants, have also been requested to check plants regularly and remove diseased or infested plants. Again, this is hard to monitor, as there is no direct contact with the gardening staff.

DETECTION AND MONITORING

Much has been written about the effectiveness of routine inspection programmes and insect trapping as an integral part of pest management. Presently, a limited inspection programme has been implemented at the South Bank building. Inspection of incoming material is the most developed aspect of this programme. The Dutton Park facility has a more comprehensive procedure in place. This is the result of a number of serious insect and mould outbreaks occurring in a short space of time, and thus placing the building in the spotlight. The poor environmental and structural condition of the building has made the facility very conducive to pest activity of all varieties.

There are three main procedures involved in the inspection and monitoring of pest activity: routine inspection of storage areas, displays and work areas; routine inspection of incoming material; and insect trapping.

Routine inspection of storage areas, displays and work areas

Currently, routine inspections only occur at Dutton Park and in identified problem areas at South Bank. The inspection programme operates at two levels. The first level involves non-conservation staff trained in basic procedures for identifying signs of pest and mould activity. A checklist is used to assist staff with the detection process. One section of the form provides space for details of any activity to be written. Staff describe what they have found and the exact location. The completed form is returned to the Senior Conservator. Staff are encouraged to collect samples if possible. Each section is provided with a 'pest kit', containing material needed for inspection and sample collection such as specimen jars and tweezers. Currently only staff involved at Dutton Park have access to these kits. Once training for South Bank staff is complete, each division within the Library will be issued with their own pest kit. The Conservation Unit will be responsible for maintaining supplies.

The second level of inspection involves the conservation staff only. Each month a thorough inspection of Dutton Park should be carried out and traps checked and replaced as necessary. A record of the catch is kept. Unfortunately due to staffing and time restraints, this level of inspection is not always possible. To try to address the situation one staff member, the conservation officer, has been made responsible for this task. Previously the whole team was involved.

Routine inspection of incoming material

An inspection procedure for checking incoming material was implemented at South Bank by the Conservation Unit in 1998. A formal procedure was introduced and all relevant staff given copies of the procedure. A quarantine area is located in the library's loading bay. Library staff place incoming collection material on the shelves provided, to await inspection by conservation staff. Due to both financial and space restraints, the quarantine area is not ideal.

The majority of units within Special Collections have been diligently sending all incoming material through the Conservation Unit, but there remains a significant gap in the quarantine system. Materials received by the cataloguing and acquisition units are amongst those that fall through the net. These two sections receive large quantities of books and periodicals from around the world on a daily basis. Much of this material goes unchecked. Any obvious signs of insect or mould activity are reported to the Conservation Unit, but more subtle activity tends to go unnoticed. At present, staff in these areas have not had pest management training. It is envisaged that once they have, they will be well equipped to identify signs of insect activity. Due to the large quantity of incoming material along with time and staff restrictions, this checking cannot realistically be done in the designated quarantine area. Within the two divisions, checking must be carried out *in situ* as part of normal duties. There are major problems with this. By the time staff get to handle material it has already been in contact with existing collections, thus the risk of widespread infestation is greatly increased. Once the

material has been processed by the acquisition and cataloguing staff, it is then distributed to the appropriate section of the library for shelving, further increasing the chance of spreading an infestation. It is hoped that any significant outbreak will be obvious enough for staff to notice it.

Another common risk occurs when the public brings material directly into the library, thus bypassing the normal inspection procedure. One way of avoiding this problem has been to ask clients to come via the loading bay so material can be checked first. This is not a problem if the client has rung before coming in. Many people turn up unexpectedly and currently there is no procedure for non-conservation staff to deal with the situation.

Deliveries of consumables have also been a source of insect activity. Often staff store the material until it is required without fully unpacking it. In the past, outbreaks of cigarette beetle and carpet beetle (*Anthrenus verbasci*) have been found in board stocks. Staff will have to check incoming consumables before taking them to the appropriate work area.

The Bindery, which adjoins the Conservation Unit, is subject to constant sources of insect infestation. Due to the large number of incoming books, initial inspection is not practical. Material from the Public Lending Division (PLD) is sent to regional libraries throughout Queensland. On return from these libraries, books requiring binding work are identified and placed on pallets in the PLD loading bay. Material can sit here for several months before being sent to South Bank. There have been a number of reports of insect activity found in this material. Cigarette beetles and booklice (*Liposcelis corrodens*) are most common. PLD staff have received basic pest training and should be checking for pest activity during their routine processing. The problem really occurs in the loading bay, which has no environmental control, and in the summer is very hot and humid, thus creating an ideal breeding ground for insect pests. Until a safe storage area can be found, this problem will be almost impossible to solve. A nitrogen chamber for anoxic treatments is proposed for the library's new building and it is hoped that material from PLD will be routinely placed in the chamber before coming into the Bindery.

Insect trapping

Monitoring is a fundamental element of a pest management programme. The use of insect traps is the key to the monitoring process. Trapping can determine the scale, type and location of an insect problem. It can also let us know if current pest control techniques are effective. The benefits of a well-organized trapping programme have long been acknowledged (Gilberg and Roach, 1991; Child and Pinniger, 1994). The main trapping method used at the library is sticky blunder traps (Agrisense 'Window' and 'Chekkit' traps).

At the time of writing, the trapping programme remains fragmented. The Dutton Park repository is presently the only building to be fully monitored. Trapping in the South Bank building has been predominantly reactive in nature. Collection staff report insect activity to the Conservation Unit and then traps are placed in the areas of concern. Often despite the best intentions, these traps are forgotten. This problem should be rectified once collection staff are appropriately trained and detailed plans of trap locations are finalized.

Staff involved in the trapping programme at Dutton Park are currently unsupervised during their routine inspections. Members of the State Reference Unit's SMART team are responsible for the majority of collections in the repository. They are expected to check for mould and insect activity as part of their daily routine. This has had some degree of success, but there is evidence that not all staff are carrying out this task and it would appear that a level of complacency has set in. To counteract this problem, conservation staff will now not only conduct monthly inspections of all areas at Dutton Park, they will also join the SMART team staff in their inspection routine. It is very difficult to maintain staff interest in such unglamorous activities.

The Conservation Unit has kept a logbook of pest and mould activity since January 1999 and all reported pest activity has been documented. On evaluating this data, no clear pattern has emerged, though there is an increase in reported pest activity during the summer months and certain parts of the library do appear to have more pest problems. Pest activity does not appear to slow down a great deal in winter as the logbook contains a steady stream of entries throughout the winter months. Unlike cooler climates with very cold, icy winters, Brisbane's winter months are relatively mild with temperatures rarely below 5°C, so many insects continue to thrive.

RESPONSE TO SIGNIFICANT OUTBREAKS

Previous responses to large-scale outbreaks have had considerable impact on the financial and staff resources of both Collection Preservation and the general divisional budget.

The cigarette beetle outbreak at the end of 1998 in the legal deposit newspaper collection provides the best example of this. This outbreak occurred before current pest management procedures had been implemented. Treatment of the whole repository was deemed necessary. This meant the treatment of approximately 3000 volumes of bound newspapers. Interestingly, the most damaged volumes had been shrink-wrapped. The creation of a microclimate conducive to cigarette beetle appeared to have been created, bringing into question the validity of routine shrink-wrapping of legal deposit newspaper volumes currently practised at the library.

Freezing was chosen as the most practical, safe, and cost effective method of eradication (Gilberg and Brokerhof,

1991). Due to the large scale of the outbreak, anoxic treatment would have proved impractical and costly. The use of chemical treatments was also dismissed because of the potential risks to human health and the possible damage caused by residual chemicals on the surface of the treated objects. In addition, it would not have killed the beetles in the shrink-wrapped volumes.

First the room was isolated and no material was allowed in or out. Blunder traps containing 'Lasiotrap' cigarette beetle pheromones were placed in all other repositories in an attempt to assess the scale of the outbreak. The traps were monitored for approximately three weeks. Fortunately, the cigarette beetles appeared to be contained in the one room. Each volume in the affected room was checked for holes in the shrink-wrapping and re-shrink-wrapped if necessary. The volumes were then packed carefully onto pallets and a final layer of plastic was applied to ensure each pallet was completely airtight. The palleted material was then transported to the P&O Cold Storage Facility usually used by the shipping industry. The pallets were frozen at −20°C for approximately two weeks over the Christmas holidays.

While material was in cold storage, the affected storage area was thoroughly cleaned, and a pest contractor sprayed the area with a pyrethrum fog. Blunder traps with the cigarette beetle pheromone were placed throughout the room to detect further beetle activity.

On return from cold storage, the material was left wrapped while thawing took place. No water damage occurred. Once thawed, the painstaking task of cleaning the volumes took place. This was done by hand using brushes and low suction vacuum cleaners. There was an extraordinary amount of damage along the spines of the volumes (Figure 3) where the larvae had eaten the adhesive, but little damage to the pages themselves. Despite trepidation from the Conservation Unit, material was re-shrink-wrapped and re-shelved.

Further isolated outbreaks of cigarette beetle have occurred since and will continue to occur until the environmental conditions in the building improve.

The issues of staff resourcing and the impact of large-scale operations added to the standard conservation programmes were highlighted. With the exception of a few casual and John Oxley staff, the clean-up team consisted predominantly of staff from the Collection Preservation branch, which consists of the Conservation Unit, Bindery, Image Production Unit and Microfilm Unit. The entire conservation team was absent from normal duties for over a month. This had a serious impact on the planned conservation programme, taking many months for the unit to return to normal. There was a disappointing lack of cooperation from other divisions. If a similar sized outbreak occurs again, a casual workforce will be trained to do the work, with only one member of conservation present to supervise. Naturally, the necessary funding will be required.

Table 2 Current pesticide usage at the State Library of Queensland

Trade name/ Manufacturer	Insecticide (other names)	Active constituents	Location used	Application frequency (months)	Comments
RESLIN® Thermal fogging & ULV insecticide concentrate	none	Piperonyl butoxide Bioresmethrin	South Bank PLD Dutton Park	3 6 3	
Magnetic roach food (MRF) cockroach bait	none	Boric acid	PLD	6	
Maxforce® professional insect control roach gel	none	Hydramethylnon	South Bank Dutton Park	3 3	
Garrard's termite powder insecticide	Arsenic trioxide dust, Termite powder	Arsenic Trioxide 50% Red Oxide Dye 50%	South Bank	3	Potential human carcinogen
Rentokil alphachloralose	none	Alphachlorose	South Bank	3	
Actellic public health insecticide	none	Pyrimiphos-methytech. Methyl isobutyl ketone	South Bank	3	Marine pollutant. South Bank is located by the Brisbane River
Baytex 550 insecticide	none	Fenthion	South Bank	3	
Rentokil super rat drink concentrate	none	Bromadiolone	South Bank	3	Very poisonous if swallowed. Native animals in area.
CISLIN® residual insecticide	K-Obio, K-Otek	Deltamethrin	South Bank PLD Dutton Park	3 6 3	
Coopex insecticidal dusting powder	none	Permethrin	South Bank PLD Dutton Park	3 6 3	
Crackdown residual insecticide	none	Deltamethrin D-tetramethrin 20:80	South Bank PLD Dutton Park	3 6 3	
Demand insecticide	none	Lambdacyhalothrin technical	South Bank	3	Marine pollutant
Ficam W® insecticide	Non manufacturer's product	Bendiocarb	South Bank PLD	3 6	
Responsar beta SC professional insecticide	none	Betacyluthrin	South Bank PLD Dutton Park	3 6 3	
Talon® rodenticide all weather wax blocks	none	Brodifacoum	South Bank PLD Dutton Park	3 6 3	
Starycide 480 larvicide	none	Triflumuron	South Bank PLD	3 6	
Vapona 500 insecticide	none	Dichlorvos	South Bank PLD	3 6	Dichlorvos is known to cause anemia in children and has neurotoxic or carcinogenic effects on animals. It is banned in a number of countries.

PEST CONTRACTORS

Having a good working relationship with the pest contractor is common sense but not always easy to achieve. Currently, routine spraying of residual insecticides continues at three-monthly intervals at Dutton Park. The contractor is not employed by the State Library but by the government agency from whom the Library rents the building. This makes it difficult to convince the contractor of the benefits of working with the Conservation Unit. Only one meeting in two years has been held with the contractor and the results of that encounter were not encouraging. The pest contractor for the South Bank and PLD buildings remains elusive although a meeting is being arranged. Spraying is carried out every three months at South Bank and every six months at PLD. A grand total of 17 different chemicals are used throughout the library (Table 2).

The whole contractor situation remains unsatisfactory and every effort is being made by the Conservation Unit to have more control over the use of unnecessary chemicals within the institution. It is hoped that when an operational IPM programme is in place throughout all library buildings, this need to rely on an armory of insecticides will end.

CONCLUSION

A cost analysis shows that almost A $30,000.00 has been spent on treating pest-infested collections in this financial year alone. An effective IPM programme would have prevented these outbreaks and saved the library a considerable amount of money. If sound economic reasoning cannot convince upper management of the value of implementing a functioning proactive pest management programme, it is difficult to know what will. The current programme remains ineffective and disjointed because there is not institution-wide support and involvement. Until this occurs and appropriate funding and staffing issues can be resolved, the programme is doomed to fail or limp on in its current form, an ineffectual but well meaning gesture.

REFERENCES

Child R, Pinniger D, 'Insect trapping in museums and historic houses', in *Proceedings of the 15th IIC International Congress; Preventive Conservation*, September 1994, Ottawa, Canada, 129–131.

Gilberg M, Brokerhof A W, 'The control of insect pests in museum collections: The effects of low temperature on *Stegobium paniceum* (Linnaeus), the drugstore beetle', in *Journal of the American Institute of Conservation*, 1991, **30**(2), 197–201.

Gilberg M, Roach A, 'The use of a commercial pheromone trap for monitoring *Lasioderma serricorne* (F.) infestations in museum collections', in *Studies in Conservation*, 1991, **36**, 243–247.

Gordh G, Law L, McGrath S, *QCC Pest Audit*, 1998, Unpublished report, Entomology Department, University of Queensland, Australia.

Ryan M, *Wildlife of Greater Brisbane*, 1995, Queensland Museum, Brisbane, Australia.

MATERIALS AND EQUIPMENT

Agrisense window trap and Agrisense Chekkit insect trap Lasiotrap®

Garrard's Pesticides Pty Ltd
32 Kenworth Place
Brendale
Brisbane QLD 4500
Australia
Tel: +61 7 3881 1693
Fax: +61 7 3881 1781

ACKNOWLEDGEMENTS

The author would like to thank all staff at the State Library of Queensland involved in the IPM programme. A particularly big thanks to the Conservation Unit: Sid Furber, Shane Bell, Brian Wilson and Tristan Koch for their tireless work. Thanks also to Reina Irmer, Leif Ekström, Garry Cranitch, Vinod Daniel and Geoff Monteith for all their help and expertise.

BIOGRAPHY

Lydia Egunnike specialized in paper and photographic conservation at Camberwell College of Arts, London 1992–94. Previously she had completed a Bachelor of Arts at Griffith University, Brisbane. Over the past ten years, Lydia has worked at a number of institutions in Australia and the United Kingdom. These included Queensland State Archives, Edinburgh University Library and University of Southampton Library. She has been Senior Conservator at the State Library of Queensland since mid-1998 and has been responsible for the implementation and operation of the library's pest management programme.

CHAPTER SEVEN

No uninvited guests: successful pest management in historic houses

Amber Xavier-Rowe and David Pinniger

COLLECTIONS CONSERVATION

English Heritage, 23 Savile Row, London W1S 2ET, United Kingdom

Tel: +44 20 7973 3324   Fax: +44 20 7973 3209   e-mail: amber.rowe@english-heritage.org.uk

Abstract

Historic houses are at risk from insect pests that cause damage to the structure of the building and the collections housed within. The most effective way of reducing the risk of pest damage is by developing an appropriate integrated pest management (IPM) programme. This programme should include monitoring for pests, modifying the environment to discourage pest attack and targeting treatment only where it is needed. Implementation of IPM may appear at first sight to be straightforward but this paper shows that application of IPM in practice means a thorough understanding of the individual houses and their contents. The range of environments, collection materials, staffing, plus the constraints and pressures of public access, must all be taken into account. This paper describes the development of IPM programmes in a number of large historic houses in the UK. It is illustrated with examples of trapping and targeted treatment, which show the need for a flexible and innovative approach. The results have shown that the importance of cleaning cannot be overestimated and this less-glamorous aspect of IPM must be given the greatest priority. It also emphasizes the key role of training and site staff involvement in any successful IPM programme.

Keywords

Integrated pest management (IPM), monitoring, trapping insect pests, housekeeping

Introduction

Historic houses, with their unique combination of architecture and accumulated collections, attract large numbers of the visiting public. They also provide an ideal habitat for insect pests, which thrive in dark, undisturbed areas commonplace in these properties. Over the past four years, English Heritage has developed an insect pest strategy, which has not only reduced the risk of damage to collections, but has resulted in improved standards of collection care. This paper introduces the strategy and describes its practical application using four case studies.

English Heritage is the leading body for the historic environment in England. It is responsible for the care and presentation of over 400 historic sites, 132 of which house collections of varying sizes. The day-to-day collections care activities are undertaken by site staff and supported by regional curators and the central Collections Conservation team. Only three houses employ historic contents cleaners, whilst the remaining properties are staffed by custodians, only part of whose jobs is collections care. Educating site staff to follow an easily implemented and maintained integrated pest management (IPM) programme is the key to reducing the risk of damage to collections. The IPM strategy developed for English Heritage properties relies on four interdependent cornerstones:

- monitoring and accurate recording of insect pests caught on sticky traps
- practical training courses followed by ongoing site support
- interpretation of the annual insect catch results from each property
- targeted action

Central coordination by the Collections Conservation team ensures that the strategy is implemented and maintained.

Monitoring

Assessing the presence of insect pests using sticky traps located in each room is a very effective early warning method (Child and Pinniger, 1994). Keeping the total number of traps down to a minimum and checking them four times a year means that site staff can readily incorporate the task into their existing workload.

One or two traps are placed in each room, usually located in the fireplaces, in which a collection is displayed or stored. The traps are examined by two site staff. This is either done in each room or the traps are gathered up and examined in one location, and then replaced. Insect pests are identified and recorded onto a standardized form. To help highlight any potential pest problems the catch results are then plotted by site staff onto house plans using coloured dots.

The easy part is laying the traps, but in order to get an accurate picture of the risks to collections, they must be regularly checked (minimum of four times a year). It is also essential that site staff identify precisely which insect pests are found on the traps.

TRAINING

The IPM training courses, run by English Heritage and led by David Pinniger, are responsible for the increase of effective monitoring programmes, from one property in 1998 to eight in 2001. Following each course, regular on-site visits further encourage the implementation of the monitoring programme and the honing of identification skills.

The English Heritage IPM course has been specifically designed to deliver practical methods that can be readily implemented in a historic house. The course is run over three days and is based in a historic property. The first half of the course outlines the risks that insect pests pose to collections and introduces the key species. An insect identification practical then helps to develop identification skills. The use of sticky traps to monitor insect pest activity is also introduced as an effective method for spotting problems and guiding targeted control.

Figure 1 Participants on the English Heritage IPM course completing the insect monitoring practical at Audley End House in Essex

The second half of the course is built around a monitoring practical, which involves small teams surveying room environments and examining a range of traps placed throughout the property (Figure 1). This helps to further develop identification skills and participants begin to understand where insects live and how the monitoring programme should be implemented. Each team reports back the catch results in a joint session, where issues relating to accurate identification and time commitments are discussed. The course is designed to encourage staff to understand that insect pest monitoring is an essential part of their job. The participants leave with a supply of insect traps, a magnifying lens, record sheets, instructions, insect pest poster and reference books.

Most participants will lay down the traps when they return to their property. However, only a few will actually start recording the catch on a quarterly basis. Therefore, follow-up site visits by members of the Collections Conservation team are crucial to the successful implementation of a monitoring programme. These visits also encourage good relationships and improve everybody's understanding of site issues relating to collections' management and conservation. Four quarterly visits are usually enough to build the confidence of site staff to a level so that they can manage the IPM programme.

Accurate identification of insect pests is crucial to the effectiveness of the IPM programmes. To improve identification skills a one-day Insect Identification Masterclass was developed. Staff who had attended the introductory three-day course and who are responsible for IPM in their property were invited. Participants practised identifying both live and dead insect specimens, using microscopes and a range of reference books. They also left with two insect identification books and an illuminated magnifier. The masterclass concept helped to promote good practice and to reconfirm the importance of insect monitoring.

A colour poster featuring key insect pests was developed as a direct result of the courses. Initially prepared to help course participants recognize pests, it was realized that this would be a valuable tool to all those involved in IPM. The poster was produced as a joint venture between English Heritage and the Museums and Galleries Commission (now Re:source) with illustrations supplied by the MAFF Central Science Laboratory.[1] To date, 4000 copies of the poster have been sent to museums and historic houses in the UK and it has now become an essential handout for all IPM courses. It has also been well received in other countries as far afield as Singapore and Sweden.

INTERPRETATION OF THE INSECT PEST CATCH RESULTS

The next part of the strategy is to interpret the catch results on an annual basis. This allows trends at a property to be assessed over a year and from year to year, to build up a

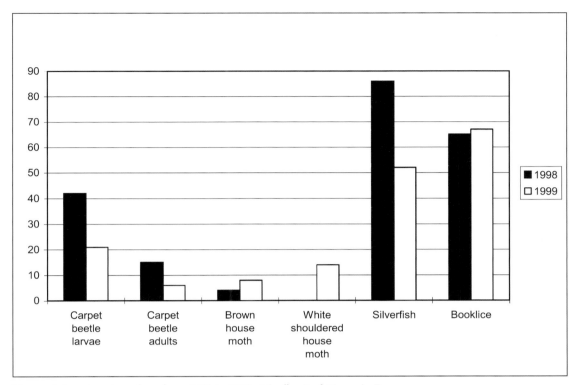

Figure 2 Insect pest numbers from 1998 to 1999 at Audley End House in Essex

clear picture of the 'hot spots'. For example, the presence of either booklice *Liposcelis* and/or silverfish *Lepisma* indicates localized damp environments. Annual assessment also ensures that insect pest management keeps a high profile and reminds all those concerned (house manager, site staff and regional management) that the data gathered from traps serve a useful purpose.

Site coordinators are responsible for recording the insect catch. The completed forms and house plans (which record the trap locations) are then forwarded to the Collections Conservation team. These results are collated to produce bar charts that clearly illustrate the catch results over the year and highlight issues that need to be followed up (Figure 2). The annual reports are sent to the site staff, the site manager and regional curator.

TARGETED CONTROL METHODS

In every case where insect pest monitoring has been implemented in a property, problems such as unstructured housekeeping, blocked chimneys and infested objects have been discovered. Removing the source of the insect pests is the first step and this usually involves unblocking and sweeping chimneys. Other sources of pests, including large deposits of flies and other dead insects, must be dealt with through targeted cleaning schedules. The Collections Conservation team at EH works with site staff and regional curators to advise and encourage them to take appropriate action.

Infested objects are treated using either freezing or heating methods, depending on the type, size and number of objects. The use of residual insecticides is limited to the selective spraying of permethrin micro-emulsion, Constrain® (Historyonics, Wales), around the edges of rooms at floor level and the spreading of a desiccant dust in inaccessible areas, e.g. chimneys. Toxic fumigants are no longer used.

CASE STUDY 1: AUDLEY END HOUSE

Audley End House, located near the town of Saffron Walden in Essex, is one of the most significant Jacobean houses in England. It was one of the great wonders of the nation when it was built by the first Earl of Suffolk, Lord Treasurer to James I. During the 18th century substantial parts of the house were demolished but what remains is still impressive. The surviving collections are displayed in over 20 rooms. Thousands of visitors are drawn to the property, as is a range of insect pests which inhabit the numerous voids inherent in a building of this nature.

The house is staffed by five full-time personnel and supplemented by 24 seasonal posts during the open season. Two part-time historic contents cleaners are responsible for housekeeping and insect pest monitoring. Annual visitor numbers are in the region of 100,000.

Audley End House was one of the first properties to be monitored for insect pests. One of the authors (Amber Xavier-Rowe) set up a monitoring programme, working with the historic contents cleaners, soon after attending the landmark course 'Pest Management and Control in Museums', run at West Dean College in October 1996. The course was organized by Re:source (formerly the Museums and Galleries Commission) and the Getty Conservation Institute.

When the traps were first checked significant numbers of woolly bear larvae *Anthrenus verbasci* were caught, particularly in the picture gallery located on the first floor. The gallery predominantly displays mounted natural history specimens, housed in glazed display cases. The insect catch results inspired a closer inspection of the room, revealing copious amounts of dead flies hidden under the base of the wool plush curtains and around the edge of the room. Live woolly bears were then spotted, embedded in the 19th century wool plush curtains behind the tiebacks, a dark undisturbed area. Damage was also evident to the modern woollen lining curtains. It appeared that the woolly bears had been attracted by the dead flies and had then moved into the curtains. But could the dead insects be the original source? The large central fireplace was then checked, and this appeared to be the prime source of woolly bears. The base of the chimney just above the hearth had been sealed with metal sheeting and staff commented that bird noises were occasionally heard emanating from this area, indicating that dead birds might be the source of the insect pests.

The author and historic contents cleaners then started examining the curtains throughout the first floor and discovered many live woolly bears. All the chimneys had been sealed just above the hearths. The carpets on the first floor were also being attacked by woolly bears, again attracted by the dead flies caught under its edges. On the second floor, which is closed to visitors and largely empty of contents, an enormous number of dead flies, ladybirds and lacewings were trapped in gaps between the windows and floors and inside window seats (Figure 3). Woolly bears feasting on the dead insects on the second floor could easily drop down to the first floor and start attacking the woollen textiles.

Following a visit by David Pinniger to confirm the hypothesis that the chimneys were a key source of insect pests, the regional curator organized for all 43 chimneys to be unsealed and cleaned. Copious amounts of debris, including pieces of carpet, dead birds and nesting materials, were removed. Infested curtains were also taken down, bagged and sent to a conservation studio for cleaning and freezing. Housekeeping activities took on a much more thorough approach to tackle the accumulated dead insects on the second floor and around the edges of the show rooms. Objects were checked and cleaned more regularly. Curtains were also drawn on a regular basis to reduce dark harbourage areas.

The chimneys and closed shutter boxes were then treated with Drione®, a desiccant dust, whilst the edges of first floor rooms were sprayed with Constrain®. This treatment was undertaken by a private pest control contractor.

The numbers of woolly bear *Anthrenus verbasci* larvae and two-spot carpet beetle *Attagenus pellio* larvae actually increased following the treatments and cleaning! This is not an unusual phenomenon, as all the extra activity had flushed them out. This proved to be the case as further trapping results showed an eventual marked decrease in woolly bears (Figure 2).

Following the discovery of insect pests at Audley End House, staff awareness and understanding of the importance of good housekeeping has dramatically improved. A targeted housekeeping programme, combined with insect pest monitoring, is keeping the numbers of

Figure 3 Large numbers of dead flies on a window sill, on the second floor at Audley End House in Essex

insect pests under control. This has benefited the long-term conservation of the collections and the building itself. Clearing the chimneys has resulted in improved ventilation with the added bonus of removing the long-term smells that permeated some of the main show rooms!

CASE STUDY 2: WALMER CASTLE

Walmer Castle, located near Dover in Kent, was built to withstand the French and Spanish following Henry VIII's break with the Roman Catholic Church. Unfortunately, the defences are not effective against repelling insect pests. It was transformed when it became the official residence of the Lords Warden of the Cinque Ports, an ancient title that originally involved control of the five most important medieval ports on the south coast. Past wardens include William Pitt the Younger, the Duke of Wellington and Sir Winston Churchill. The current Lord Warden is Her Majesty Queen Elizabeth, the Queen Mother, who resides at the site for a few days each year.

Walmer Castle is staffed by three full-time and four seasonal custodians, who have to cope with 60,000 visitors each year. Staff are responsible both for operating the

Figure 4 The large amount of rubbish removed from the chimney in the Wellington Museum room at Walmer Castle in Kent

property and caring for the collections. Following the English Heritage IPM course, one of the authors (Amber Xavier-Rowe) started working with two site staff. Sticky traps were placed in show rooms and dates for quarterly checks were agreed.

Insect pest trapping at Walmer Castle revealed enormous numbers of larvae and adults of the case-bearing clothes moth *Tinea pellionella*, mainly in the fireplace in the Duke of Wellington's bedroom. Their source was almost certainly the chimney. Staff agreed with the recommendations that all the chimneys should be swept and that the housekeeping should become more targeted. The house manager argued the case with regional management and eventually secured the funding.

When the chimneys were swept, enormous amounts of accumulated rubbish, including numerous dead birds, pieces of carpet and bird nests, were removed (Figure 4). It appeared that they had not been cleaned for over 20 years. Regional building maintenance schedules had tended to focus on essential works to the building fabric, thus overlooking perceived low-priority activities like chimney sweeping.

Insect monitoring at Walmer Castle revealed major risks to the collections and the building. The blocked chimneys and untargeted housekeeping practices had led to a large number of insect pests with resulting damage to the collections. Clearing the chimneys has also improved ventilation, which has helped to preserve the building fabric. Site staff are now convinced of the value of the insect monitoring programme and have taken practical steps to improve housekeeping and building maintenance.

CASE STUDY 3: DOWN HOUSE

Charles and Emma Darwin moved to Down House in 1842. In the Old Study, Darwin wrote one of the most significant books of recent centuries, *On the Origin of Species by Means of Natural Selection*. Down House is located in a rural location near Bromley in South London. In 1997, English Heritage completed the extensive conservation of the building and the redisplay of the collections.

Following the conservation project, high numbers of booklice *Liposcelis* were being caught on the sticky traps, particularly in the Old Study. Booklice will surface graze damp paper and textiles and only breed in an environment with humidity between 60% and 80% (Busvine, 1980). These results indicated that the building fabric was still saturated, as the house had been open to the environment during building works. To ensure that the booklice did not spread to the books in Darwin's library, spaced stacking and regular cleaning and checking routines were instigated. Ongoing monitoring over the past two years has confirmed a marked decrease in the numbers of booklice, indicating that the building fabric has dried out. Importantly, there has been no observed damage to the books and paper-based items in the Old Study.

Insect pest monitoring also led to the discovery of two-spot carpet beetle *Attagenus pellio* larvae and varied carpet beetle larvae *Anthrenus verbasci* feasting on woollen felt druggets. The druggets were laid around the edge of each room as part of the restoration of the ground floor, which was completed in 1997. Along the edges of these rooms, the drugget had been folded and tacked in position creating a dark undisturbed habitat, ideal for carpet beetle larvae. With hindsight, a synthetic material would have been preferable to wool. This is one of the first documented cases of serious damage to woollen textiles by *Attagenus pellio*.

The discovery of insect pests in the woollen felt druggets generated a whole new approach to housekeeping at Down House, where each member of the custodial team was assigned one ground floor room and its contents to clean. This approach has generated a strong sense of ownership and responsibility for the care of 'their' room. It has provided a model for other English Heritage properties that rely on custodial staff for the day-to-day care of the collections. Ongoing results from the insect traps have confirmed that the cleaning and localized treatment with Constrain® has brought the carpet beetle larvae under control at Down House. Because of the early warning of an *Attagenus* infestation and the prompt action taken, damage to historic objects was prevented.

CASE STUDY 4: BRODSWORTH HALL

One of England's most complete Victorian country houses, Brodsworth Hall, near Doncaster in South Yorkshire, opened in 1995 following a major programme of conservation by English Heritage. Built in the Italianate style and decorated and furnished in the opulent fashion of the 1860s, much of the original collections survive intact.

At Brodsworth Hall, pest control methods initially involved chemical and freezing treatments and, more recently, the use of heat. When English Heritage acquired the contents of Brodsworth Hall in 1990, most of the rooms had been undisturbed for many years, providing an ideal environment for insect pests. Natural history specimens had been completely stripped to the skeleton by woolly bear larvae *Anthrenus* sp. and webbing clothes moth larvae *Tineola bisselliella*. All the carpets and textiles were frozen on-site in a mobile unit (Berkouwer, 1994), whilst other objects were bulk-treated by a pest control contractor using methyl bromide.

An IPM programme started at Brodsworth soon after opening. It consisted of over 100 traps, which were checked on a weekly basis by two historic contents cleaners. The insect traps could only be superficially examined due to their large number and the requirement to check them weekly. Small larvae and booklice were not being recorded. We have found that when trapping results include booklice, it is an indication that site staff are closely examining the traps. Following training and on-site support, the IPM programme at Brodsworth was reduced to 60 traps, examined four times a year. The results have provided a much more accurate indication of the risks from insect pests.

Monitoring over the last four years by the two historic contents cleaners has shown a low level of pest activity. Any activity was mainly booklice *Liposcelis* and silverfish *Lepisma* with the golden spider beetle *Niptus hololeucus* living in chimneys. A few woolly bear larvae *Anthrenus*

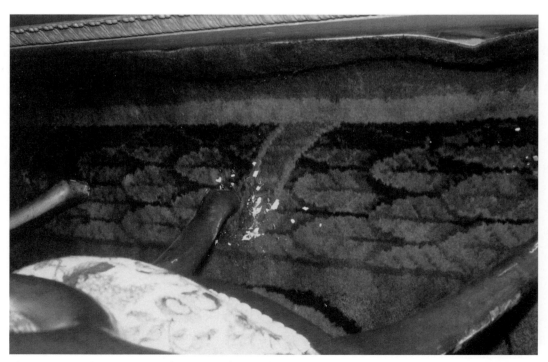

Figure 5 Mice damage to an 18th century gilt side chair located in the drawing room at Brodsworth Hall in South Yorkshire

have recently reappeared in local areas. Exemplary housekeeping practices at Brodsworth Hall are largely responsible for the low level of pest activity.

Furniture stored in the stable block was recently found to be infested with woodworm *Anobium punctatum* and deathwatch beetle *Xestobium rufovillosum*. A temporary chamber, constructed from plywood and heated by wall-mounted forced fan units, was successfully used to treat the collections (Xavier-Rowe *et al.,* 2000). Heat has a number of advantages over other methods, as objects do not require bagging and the treatment time is considerably faster than freezing or low oxygen atmospheres.

During the 1999 winter closed-season, damage was noticed to gilt side chair legs, which was caused by mice gnawing on them (Figure 5). The house staff initially laid 'break back' traps and caught two mice. They then set up a mouse-monitoring programme throughout the ground floor using humane traps, which were checked weekly. No further mice have been trapped and no further damage has been observed.

CONCLUSION

Successful pest management in historic houses is based on the delegation of responsibility for IPM to staff based in the properties. This continues to be achieved at English Heritage through a strategy which promotes an achievable programme of IPM, supported through practical training courses, regular on-site visits and central coordination.

ENDNOTES

[1] The poster 'Insect pests found in historic houses and museums', is available free of charge from English Heritage. Product code XH20135. Tel: +44 870 333 1181 e-mail: customers@english-heritage.org.uk

REFERENCES

Berkouwer M, 'Freezing to eradicate insect pests in textiles at Brodsworth Hall', in *The Conservator*, 1994, **18**, 5–22.

Busvine J R, *Insects and Hygiene*, 1980, Chapman and Hall, London.

Child R E, Pinniger D B, 'Insect trapping in museums and historic houses', in *Preventive Conservation, Practice, Theory and Research*, Roy A and Smith P (Editors), 1994, IIC, London, 129–131.

Xavier-Rowe A, Imison D, Knight B, Pinniger D, 'Using heat to kill museum insect pests – is it practical and safe?', in *Tradition and Innovation Advances in Conservation*, Roy A and Smith P (Editors), 2000 IIC, London, 206–211.

MATERIALS AND EQUIPMENT

Constrain® (water-based permethrin micro-emulsion), museum traps, and KIO® (Drione®, desiccant dust is no longer available in the UK. A similar product, KIO®, is available from Historyonics).

Historyonics
17 Talbot Street
Pontcanna
Cardiff CF11 9BW
Wales
Tel: +44 29 2039 8943
Fax: +44 29 2021 8672

BIOGRAPHIES

Amber Xavier-Rowe developed her interest in the conservation and management of collections in the historic house context since joining English Heritage in 1995. Prior to that she worked at the Leather Conservation Centre and Artlab in Australia. She originally trained in the conservation of objects at the University of Canberra. She is an accredited member of the United Kingdom Institute for Conservation.

David Pinniger is an entomologist who provides consultancy and training on pest management. He was a research entomologist at the MAFF Central Science Laboratory in Slough and is now an independent consultant advising museums, historic properties, local authorities and companies providing pest control services. He is the pest management strategy adviser for English Heritage and many of the major UK museums.

CHAPTER EIGHT

TRAINING FOR MUSEUM STAFF IS A PREREQUISITE FOR SUCCESSFUL INSECT PEST MANAGEMENT

Valerie Blyth

SENIOR TEXTILE CONSERVATOR

Textiles Section, Conservation Department, Victoria and Albert Museum, Cromwell Road, South Kensington, London SW7 2RL, United Kingdom

Tel: +44 20 7942 2123   e-mail: valblyth@vam.ac.uk

ABSTRACT

The success of the insect pest management strategy in the Victoria & Albert Museum (V & A) has been due to the high level of staff training. The first written guidelines and procedures were produced in 1989, following an outbreak of carpet beetle infestation in the museum. The training and implementation of pest management has been conservation-led and the museum policy on insect pests was written by conservation staff. Initial training was given by a pest control consultant entomologist, David Pinniger. Training for conservators and curators covers: the basic identification of main insect pests; use of sticky blunder traps; an understanding of the damage pests can cause; and which materials are vulnerable within the museum collections. Training includes a survey of the V & A environment, including old buildings, roof spaces, pigeon roosts, vents, grilles and dead spaces. Training a wide range of museum staff has been a priority, to heighten awareness of the insect pest problem and to ensure that staff can report back accurately. There has been a huge raising of awareness of all insect pest issues throughout the museum as a result of the insect pest management programme. A pest management training manual has now been produced for staff. This will reach a wider audience by being available on both the V & A intranet and the V & A museum website. Our experience shows that the development of a successful integrated strategy and the development of insect pest training go hand in hand.

KEYWORDS

Museum pests, pest management, training

INTRODUCTION

Insects are persistent, thus requiring staff perseverance in the continuing war that has led to the high level of insect pest awareness at the Victoria and Albert Museum (V & A). How we have achieved this and how the training of museum staff has been central to the success of the pest strategy of the museum, will be discussed. This paper reviews the history of pest training needs in the V & A and gives examples from the training manual co-authored with David Pinniger.

A 'bug committee' has met regularly since 1989 to discuss insect pest issues. As the current chair, I collated contributions from other members in a report detailing the work of the committee. This report is for the central management team to ensure continuing funding for pest control.

David Pinniger and I originally developed a 'bug and pest control workshop' as part of the Grade G Development programme (now the Curatorial Assistant Development programme). The workshops have formed an essential part of the programme and will continue to run as the scheme evolves. To date, five workshops have been held since 1996 and these have been attended by a total of 47 staff. Feedback has been consistently excellent with the majority of ratings being four or five (five being the highest score). We have been responsive to feedback from participants and have continually adapted the programme to meet changing needs. A copy of the training manual is given to each participant at these workshops and, as a response to requests, now includes colour pictures of the main pests with size scale bars.

Some of those staff trained will then participate in the work of the bug committee, either as members or as those responsible for the setting and regular monitoring of insect traps. They will all certainly come across insect pests again. This may be by an occasional sighting of an insect, or when incoming loans for exhibitions are discovered to be infested. They will also use their training when assessing new acquisitions and objects brought into the museum.

The training of staff has proved invaluable to ensure that the correct procedures have been followed and the appropriate staff alerted. There was near panic when a wasps' nest formed part of a case display in the National Art Library. Inside the case there was a liberal sprinkling of dead wasps, and warding staff noticed live 'woolly bears' crawling out of the wasp carcasses. Warding staff immediately notified the appropriate staff. It was established later that there were several species of carpet beetle larvae, one of which was not normally resident in the museum. As there were carpets in the gallery near to the case containing the wasps' nest, the display was removed as soon as possible. The importance of training was again proven when warding staff alerted curatorial staff to insect activity in a packaging exhibition in the 20th century gallery. Staff checked the contents of the exhibition case and found adult and larval moth activity, later identified as *Plodia interpunctella* (Indian meal moth), in items of dried food. All infested items were treated by freezing to a temperature of −30°C.

Preventive conservation courses, which include insect pest control, are also taught to a range of staff including: students attending MA courses; curatorial assistants; and new members of staff at induction. Over the years there have been additional training sessions in galleries to demonstrate the most effective placement of the sticky insect traps used to monitor the insect populations.

The training of specialist and non-specialist staff has been crucial to the success of pest management in the museum. Training sessions have been held for museum cleaners and object cleaners, resulting in a marked improvement shown by reduced trap catches in galleries that had been thoroughly cleaned. A training session in 1993 was attended by 60 cleaners and 155 warding staff attended the same session later the same year. Warding staff training continued up to 1998 when the security and warding system was restructured. Recently, training of cleaning and warding staff has been difficult to implement. The problems are mainly logistical, as it is difficult to get staff together at the same time for training when they are working shifts. The training of smaller groups is too labour-intensive for the trainers. Recent restructuring of the former Buildings and Estates Department into the new Facilities Management Division, combined with the contracting out of cleaning and warding staff, has meant an increase of staff working in the museum and a quicker turnover of staff. This has made training of all new staff impractical. The bug committee members have put forward a variety of suggestions for how this problem might be addressed. The production of a training video has been high on the agenda for a number of years, but funding has not been forthcoming. The V & A museum has introduced an Intranet for museum staff where insect pest information

will be placed. This will make pest awareness training accessible to more staff.

Extracts from the current training manual are reproduced below and include the aims of the course and the V & A museum policy on insect pest management.

AIMS OF THE WORKSHOP

The aims of the workshop were to enable staff to:

- identify the main insect pests, the objects at risk from them and the signs of damage
- understand the life cycles of pests and what they eat
- identify environmental factors which encourage pests
- understand the need for monitoring and the way trap results are used
- know the correct course of action to take when pests are found

V & A POLICY ON PESTS

All collections are at risk and may harbour risks for other parts of the collection. At particular risk are the organic materials such as textiles, costumes, furniture, ephemera and ethnographic collections. This policy does not cover micro-organisms or larger vermin such as mice or rats. The policy of the museum is given below:

- it is the responsibility of *all* staff to be alert and aware of the threat to objects posed by insects and pests
- appropriate training and support will be given to staff on matters of insect and pest control
- active monitoring programmes will provide information on activity levels and locations
- a level of insect and pest activity across the estate will be tolerated
- remedial action will be taken as and when the activity level poses a significant threat to objects or collections
- there will be a rolling programme of controls that will include object treatments and building treatments
- particular regard will be paid to objects entering the estate, whether they be V & A objects or objects on loan
- the treatments used to control and monitor insect pest activity will be studied
- materials and fabrics used in the museum for decorative purposes should be chosen with care so as not to be food sources for insects and pests

INTRODUCTION TO PEST MANAGEMENT

The key stages are:

- recognizing and identifying priorities for action
- identifying responsible staff
- taking action on the high priorities
- establishing procedures for forward planning, financing and review

In order to develop an insect pest management strategy it is important to understand and recognize some of the key components of successful pest control. These are:

- avoiding pests – by keeping pests out
- preventing pests – by denying them safe haven
- recognizing pests – the main species and the damage they cause
- assessing the problem – by inspection and trapping
- solving pest problems – by improving the environment and carrying out appropriate treatments
- reviewing insect pest management procedures periodically and changing when necessary to improve the strategy

WHAT ARE THE PESTS?

Most insect pest problems in the V & A collections are caused by the Guernsey carpet beetle *Anthrenus sarnicus* (Figure 1) and the brown carpet beetle (more commonly known as the vodka beetle) *Attagenus smirnovi*. Other pests such as clothes moths *Tineola* and biscuit beetle *Stegobium* are also found in the museum. Other insects such as booklice and silverfish can also cause damage or nuisance. A list of some of the common insect pests and the damage they cause is given below:

- Woodworm/furniture beetle *Anobium punctatum*. Produces small round exit holes, gritty frass in tunnels. Damages sapwood of hardwoods, softwood, plywood with animal glue, composite cellulose materials and books.
- Carpet beetle *Anthrenus* sp. Produces irregular holes in textiles, loose fur, short hairy cast skins of larvae. Damages wool textiles, fur and feathers. Eats dead insects.

- Carpet/fur beetle *Attagenus* sp. Produces irregular holes in textiles, loose fur, long banded cast skins of larvae. Damages wool textiles, fur and feathers. Eats dead insects.
- Webbing clothes moth *Tineola bisselliella*. Produces large irregular holes with quantities of silk webbing tubes and gritty frass. Damages wool, fur and feather textiles, bird and mammal skins.
- Case-bearing clothes moth *Tinea pellionella*. Produces irregular holes and grazed fabric with loose silk bags. Damages wool, fur and feather textiles, bird and mammal skins.
- Biscuit beetle *Stegobium paniceum*. Produces round exit holes and gritty dust. Eats dried food and spices, starchy plant specimens, seed heads, papier mâché and freeze-dried animal specimens.
- Spider beetle *Ptinus* sp. Produces some holes or cavities with spherical silk pupal bags. Eats starchy dried plant specimens, seed heads and animal specimens.
- Booklouse *Liposcelis* sp. Produces scratched and eroded surface of materials. Eats starchy paper and glues.
- Silverfish *Lepisma* sp. Produces irregular scratched and eroded surface of materials. Eats animal glue, damp paper and textiles.

The key to avoiding pests is understanding what makes them thrive and increase in numbers. The conditions found in the V & A museum and stores are too dry for the early stages of furniture beetle or woodworm. Active infestation is only found on wood or objects which are brought into the museum from storage where humidities rise above 60% relative humidity. Pests such as silverfish and booklice also require damp conditions and

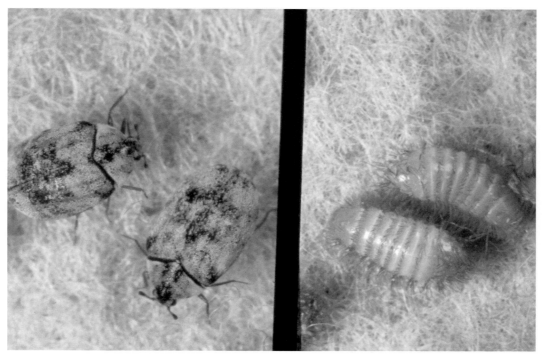

Figure 1 Guernsey carpet beetle adult and larvae (woolly bear), *Anthrenus sarnicus*

therefore by denying pests food, warmth, humidity and harbourage, we can prevent them becoming established and causing damage.

MUSEUM OBJECTS AT RISK FROM INSECT ATTACK

The museum houses a variety of collections; some are more vulnerable to insect attack than others.

New acquisitions and loans are particular sources of risk

The most likely items to contain sources of food for insects are those which contain animal skins, fur, feathers, hair, parchment, vellum and wool. Silk, linen and cotton can be attacked by pests when soiled. Other items containing sapwood, plywood with animal glue, some composite objects containing cellulose, starchy paper and papier mâché are also at risk.

Examples of collections at risk

Some examples of collections which are at risk are given below:

- *Textiles and dress.* Costume, carpets, tapestries, accessories and trimmings.
- *Furniture and woodwork.* Textile upholstery, horse hair, gilded leather panels, felt inside cabinet drawers, etc.
- *Far Eastern.* Carpets, costumes and accessories.
- *Indian.* Costumes, carpets and items with leather.
- *Theatre Museum.* Costumes and puppets.
- *Bethnal Green Museum of Childhood.* Costumes and toys.
- *Wellington Museum.* Costumes and furnishings.

Any vulnerable objects in all V & A stores on all sites

Some materials such as clean cotton are not attacked unless, for example, insects eat through a cotton covering when they emerge from infested upholstery. Paper is rarely attacked unless it is dirty and damp. In general, dirty and neglected objects in dark places will be more at risk than those that are clean and in well-lit areas.

INSECTS IN THE BUILDING ENVIRONMENT

Insects will be flying around in the summer months and some measures can be taken to stop them from gaining direct access to the building. Doors and windows can be fitted with unobtrusive sealing barrier strips to prevent the entry of larger insects. Some windows and doors can be fitted with fly mesh screening, but this is usually not possible on any outside windows at the V & A because it is aesthetically unacceptable.

Food and harbourage

Because insects are small they may find sufficient food in relatively small areas which may not be immediately obvious. The most common sources of insect problems are:

- old bird, wasp and bee nests in attics

- old heating and ventilation ducts
- cavity walls and floors
- unused rooms or cupboards, particularly in attics and basements
- gaps between walls and floors
- dead spaces behind and under storage cabinets, display cases and plinths
- dead spaces under and behind storage shelving
- felt lining on boxes and felt sealing strips on doors
- old and discarded display material, particularly when covered by wool felt
- deaccessioned material which has not been removed

Insects will penetrate small cracks and crevices. However, display and storage furniture can act as a further barrier to pest attack if it is well designed and maintained. Cupboards, cabinets and drawers which may appear to be sound should be inspected because they may have hidden cracks which allow insect access.

Temperature

Warm temperatures of 20°C and above will encourage insects to breed and so cool conditions are recommended. Although it may not be possible to lower temperatures in public areas, storage areas should be operated at as low a temperature as possible. Direct sunlight can cause localized hot spots even in cool areas and uneven temperatures may result in localized condensation.

Humidity

Many insects, such as biscuit beetles, will survive at low humidities, but some species thrive when it is damper. The number of furniture beetles or woodworm has declined in recent years due to the increased use of central heating, which has reduced average humidity levels. This pest will only successfully complete its life cycle when wood is in an environment above 60% relative humidity. It is usually only found infesting wood in basements, attics and objects that have been stored in outbuildings. Silverfish will breed rapidly and cause serious problems only in conditions of above 70% relative humidity. Booklice also need higher levels of humidity than are normally found in libraries and archives. They are often found in damp basements or in localized damp areas. Relative humidity levels should be measured and monitored and sources of dampness such as condensation, poor damp-proofing or leaks from gutters or water pipes, should be checked and repaired.

INSECT MONITORING AND RECORDING USING INSECT TRAPS AND TRAP CATCHES

Insect monitoring using insect traps is the key component of insect pest management in the V & A museum. Traps are placed at strategic points around the museum and a record kept of the number and types of insects caught.

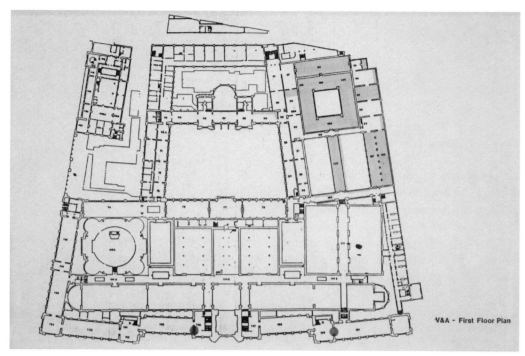

Figure 2 Plan of museum showing the focus of the Guernsey carpet beetle infestation in shaded areas

Use of traps

Traps are used to detect insects, not to control them. Sticky traps work on the principle of the wandering insect blundering into the trap and becoming stuck on the non-toxic adhesive surface. Further information about traps is given below:

- Traps should be placed in a regular grid pattern and all traps labelled with the date and their position marked on a plan.
- Traps are designed to be placed on the floor and they work best in corners and wall or floor angles rather than in the middle of open areas.
- Most traps will remain effective for at least a year.
- Traps should be checked at regular intervals. The minimum frequency of trap checking is four times a year (March, June, September and December). Traps in collections with more vulnerable material are checked monthly. In areas where serious problems have occurred, the traps should be checked at intervals of two weeks.
- The more traps used the greater the chance of finding insects, but the workload should not be underestimated and trapping programmes should be designed to be manageable.
- Insect pests caught on traps should be identified and recorded. It is important to distinguish between the two main species of carpet beetle (*Anthrenus* and *Attagenus*). It is also important to record whether the insect is an adult beetle or a larva.
- Sometimes large numbers of non-pest insects are caught on traps if they are near an outside door. When this happens, the traps should be replaced more frequently

or the trapped insects will become food for pests.
- One trap catch may not mean much – it is only by recording results over a period of time that a picture will emerge. Trap results are used as a supplement to visual inspection and the information is used to target preventative and remedial measures (Figure 2).

LOOKING FOR INSECTS: THE BUG HUNT

Look for insects in dark areas using a good flashlight. Look inside folded textiles or where they are touching walls or floors. Look for signs of insect activity such as fresh frass from woodworm exit holes, moth webbing or cast skins of carpet beetle larvae (woolly bears). Examine window sills and light fittings in spring and summer for signs of adult insects.

Even a thorough search or inspection may fail to find insects which are hidden away. This is why the V & A museum has a programme of monitoring using sticky traps to detect and find insects.

Before you decide on priorities and any action that is needed, it is useful to go through a checklist:

- Are there pest insects on traps?
- Is there damage to objects?
- Are there signs of insects on objects?
- Are they alive or dead?
- What species are they?
- How many insects are there?
- Are they breeding?
- How many objects are affected?
- Are they in display material?
- Are they elsewhere in the building?

PREVENTION OF PESTS: CLEANING AND QUARANTINE

Cleaning

Cleaning is the most important part of any insect pest management programme. Many cleaning schedules are targeted on the most obvious public areas, which may appear to be superficially clean. A close examination with a good flashlight will usually show accumulations of organic dirt and debris in corners, wall or floor angles and behind fittings that will support insect pests. Unused rooms and storage areas are often neglected and dirt and debris will provide an ideal harbourage for insects.

Quarantine

An essential part of any pest prevention policy in a museum is to keep pests out of collections. Insects can be introduced from many sources including new acquisitions, objects on loan from other museums and objects returned from loan.

Objects must be checked for infestation before being allowed into the main collection area, whether in store or on display. Inspection may reveal insect damage and clothes moth webbing, but insect eggs or small larvae may be difficult to see. Woodborer emergence holes may be obvious but developing larvae will be hidden in the wood. This means that some incubation period may be necessary to determine whether an infestation is active or dead.

A summary of the quarantine requirements and procedures are as follows:

- *Inspection area.* All organic material and objects coming into the museum must be placed in a designated area and checked for signs of infestation. If there is active infestation then objects should be treated as soon as possible.
- *Holding area.* If there is no direct evidence of live insects but there is reason to suspect that there may be active infestation, then objects should be isolated and incubated over a summer period. If adult insects are seen to emerge then remedial action can be taken.
- *Treatment area.* Treatment which can be carried out on-site by museum staff includes freezing to temperatures of −18°C or −30°C. Treatment with nitrogen or carbon dioxide may need a special area or may have to be carried out off-site by specialist staff.

All staff involved with the handling and movement of objects must be aware of the strategy and must abide by it. Any breach can result in infestation being introduced undetected into collections where they may thrive and cause damage.

PHYSICAL REMOVAL, FREEZING AND CHEMICAL TREATMENT

Prevention is better than cure and all the points made in the section on preventing pests should be the first priority.

However, if pests are found in objects or in the building then some remedial action may be necessary:

- isolate any objects suspected of being infested to prevent spread of infestation to other objects
- clean infested areas and destroy insect bodies and debris
- decide on the most appropriate treatment for the object and environment

This course booklet is not intended to provide an exhaustive review of all possible treatments but to provide a summary of the control options available at the V & A museum.

Insects in the environment

Targeted treatment using an approved insecticidal dust or spray can be very effective in reducing the numbers of insects. Insects, which live in harbourages, can only be killed when they wander across treated surfaces. Types of treatments are given below:

- wall or floor angles can be treated with a residual insecticide such as Empire 20 (encapsulated chlorpyriphos)
- dead spaces in ducts and under cabinets can be treated with an insecticidal or desiccant dust such as Drione® or Eatons KIO®
- infested textiles such as carpets and wool felt underlays, which are part of displays and not historic objects, can be treated *in situ* with an appropriate insecticide such as Constrain®
- aerosols or airborne sprays must not be used as they achieve little control of most museum pests and cause indiscriminate contamination of objects

Many localized treatments can be safely carried out by museum staff but large-scale treatments are carried out by the museum's contractor.

Insects in objects

The choice of remedial treatment will depend upon the severity of the infestation, the type of material and the value of the object. Treatment of objects should only be carried out after checking with a conservator or collections care specialist. Some examples of remedial treatments are given below:

- *Low temperature.* Objects must be sealed in bags and exposed to −30°C for three days or to −18°C for at least 14 days. It is not necessary to treat objects twice at −18°C. A chest freezer is more efficient than an upright freezer. Freezer temperatures should be checked with a thermometer. Objects should not be unbagged until they have reached room temperature.
- *High temperature.* Objects can be treated without bagging in a special humidity-controlled chamber at 52°C in the Thermo Lignum® process (Child, 1994).
- *Carbon dioxide.* Treatments at 60% concentration are usually carried out by a contractor using a special gas-

tight bubble. This has no deleterious effect on objects at normal temperatures and humidities, but long exposures of three weeks or more may be needed to kill pests.

- *Nitrogen.* Nitrogen kills insects by excluding oxygen and therefore only works at very high concentrations of 99% and above. Treatment must be carried out in specially constructed chambers or cubicles made from oxygen barrier film. The nitrogen must be humidified and oxygen levels must be very carefully monitored and controlled using an oxygen meter. As with carbon dioxide, exposures need to be longer at lower temperatures. Smaller objects can be treated successfully by sealing them in a sealed oxygen barrier film bag with an oxygen scavenger such as Ageless®.

All pest treatments should be recorded where possible.

CONCLUSIONS

The V & A museum's integrated strategy for pest prevention has been successful in preventing insect damage to objects in the collections. A low level of insect pest activity is tolerated in the museum but action is taken when insect levels pose a significant threat. The monitoring programme provides the basis for decisions on cleaning and targeted treatment of objects in the museum environment. To develop this successful integrated strategy, our experience shows that an awareness of insect pests and training go hand in hand.

REFERENCES

Child R E, 'The Thermo Lignum® process for insect pest control', in *Paper Conservation News*, 1994, **72**, 9.

MATERIALS AND EQUIPMENT

Insect traps and Constrain® (water-based insecticide, neutral pH, 0.2% permethrin)
    Historyonics
    17 Talbot Street
    Pontcanna
    Cardiff CF11 9BW
    Wales
    Tel: +44 29 2039 8943
    Fax: +44 29 2021 8672

Empire 20 (encapsulated chlorpyriphos)
    Contractor: Cannon Hygiene Ltd
    Environmental Services Division
    Unit 2
    Sovereign House
    Sovereign Park
    Coronation Road
    Park Royal
    London NW12 7QP
    United Kingdom

Drione® (1% w/w natural pyrethrins + amorphous silica gel)
    Killgerm Chemicals Ltd
    Denholme Drive
    Ossett
    West Yorkshire WF5 9NB
    United Kingdom
    Tel: +44 1924 277631
    Fax: +44 1924 264757

Eatons KIO® system (silicone dioxide or diatomaceous earth)
    Industrial Pesticides
    7–29 Brasenose Road
    Liverpool L20 8HL
    United KIngdom
    Tel: +44 151 9337292
    Fax: +44 151 9223733

Ageless® (oxygen scavenger containing powdered iron oxide)
    Conservation by Design Ltd
    Timecare Works
    5 Singer Way
    Woburn Road Ind. Estate
    Kempston
    Bedford MK42 7AW
    United Kingdom
    Tel: +44 1234 853 555
    Fax: +44 1234 852 334

ACKNOWLEDGEMENTS

Jonathan Ashley Smith, Head of Conservation, Timothy Carpenter and members of the V & A 'bug committee'.

BIOGRAPHY

Val Blyth has been a senior conservator with the V & A since 1988 where her specialisms are tapestry and carpet conservation. Her interests also include pest control, targeted treatments including major freezing projects, chairing the museum bug committee, staff training and advising colleagues on pest control in the museum.

CHAPTER NINE

Trapping Used in a Large Store to Target Cleaning and Treatment

Helen Kingsley and David Pinniger

Science Museum, Exhibition Road, London SW7 2DD, United Kingdom
Tel: +44 20 7602 1397   Fax: +44 20 7603 3498   e-mail: hkingsley@nmsi.ac.uk

Abstract

Insect traps have been used to monitor and provide early warning of pest presence in the Science Museum's small objects store. Insect traps were first used in late 1993 when an endemic infestation of carpet beetles was discovered in the building while objects were being moved from another site. The results have been analysed for the six-year period from 1994 to 2000. The major pest species in the building are the Guernsey carpet beetle *Anthrenus sarnicus* and brown carpet beetle (or vodka beetle) *Attagenus smirnovi*. The biscuit beetle *Stegobium paniceum* has also been found in numbers in specific areas. Other species trapped include the spider beetle, hide beetle, wood weevil, webbing clothes moth, silverfish and booklouse. The trapping programme has worked well to identify problem areas, which have then been cleaned and treated to prevent pests spreading into objects. This targeted cleaning and treatment has resulted in minimal damage to objects with the most effective use of limited resources.

Keywords

Carpet beetle, biscuit beetle, insect traps, target cleaning, deep clean, desiccant dust, environmental conditions, integrated pest management (IPM)

Introduction

The Science Museum's small objects store, located at Blythe House in west London, was built at the end of the 19th century and was used as the Post Office Savings Bank headquarters until the late 1970s. During the 1980s, it was converted to a storage facility by the three national museums that now share it: the British Museum, the Victoria and Albert Museum and the Science Museum.

The Science Museum occupies approximately one third of the building, occupying 9000 m² of floor space. This consists of one hundred storerooms on six floors plus four mezzanine levels, which house approximately 250,000 objects. The collections are mixed media ranging from scientific instruments, transport models, and hand and machine tools to ethnographic material. The objects are stored on open racks, in glazed cupboards or are free standing.

The first stage of integrated pest management (IPM) is the use of insect traps for monitoring insect activity (Pinniger *et al.*, 1998). We therefore placed sticky insect traps throughout the store in 1994. Eighty-five traps were initially put down and this number has been increased over the years to 180 as a response to movement of objects into the store and discovery of infestation. The traps are checked four times a year, the insects are identified and the results recorded on trap data sheets. We have found that it is

essential to use a microscope to identify newly hatched carpet beetle larvae on traps.

Problem pests

The major pest species in the building are the Guernsey carpet beetle, *Anthrenus sarnicus*, and the brown carpet beetle (or vodka beetle), *Attagenus smirnovi*. Biscuit beetles, *Stegobium paniceum*, have also been found in large numbers in specific areas.

The Guernsey carpet beetle life cycle from egg to larva to adult takes approximately one year, although it can be quicker depending on environmental conditions of temperature, relative humidity and food availability (Coombes and Woodroffe, 1983). Adults fly readily and are attracted to north-facing windows. The larvae eat proteinaceous material including dead insects, and will damage wool and textiles. The larvae, which are often known as woolly bears, are very small when they first hatch but grow to a length of about 5 mm and are very mobile.

The brown carpet beetle life cycle generally takes approximately one year but this can be either shortened or lengthened depending on the environmental conditions (Dvoriashina, 1988). Adults fly freely and are attracted to tungsten lights. The larvae are very mobile and will attack both proteinaceous and starchy cellulose materials.

The biscuit beetle life cycle takes between four months to one year depending on environmental conditions (Lefkovitch, 1967). Adults only fly at temperatures above 22°C. The larvae, which are not mobile and are usually hidden in the food source, attack a wide variety of foodstuffs but prefer starchy material.

RESULTS OF TRAPPING

The first full year of the trapping programme in 1994 showed that many Guernsey carpet beetle adults and larvae and biscuit beetle adults were found in the traps, and on the floors. In contrast, there was very little brown carpet beetle activity. Very few adults were caught and even fewer larvae, which showed that they were not established at that time. The data collected over the years indicate that the distribution of the three main insect pests is as follows. Brown carpet beetle adults are predominately found on the second floor, whereas the larvae are located in the basement and second floor rooms. Guernsey carpet beetle adults are predominately found on the fourth floor, whereas the larvae are found throughout the building. Biscuit beetle adults are found mainly on the fourth floor and basement rooms.

Data also revealed that the site did not have an infestation problem with brown carpet beetles until the end of 1994 (Figure 1). The infestation was probably a direct result of the movement of the chemistry collections from the South Kensington Museum in late 1993 and early 1994. At that time, South Kensington had a major infestation problem with brown carpet beetles. The objects were transported to the store where they were immediately placed in their storage rooms on the second floor. At that time, there was no quarantine area set up and there was no inspection of objects for pests, either before they were packed or when they arrived at the store.

Other species which have been found on traps include spider beetles *Ptinus tectus* and *Niptus hololeucus,* hide beetles *Dermestes* sp., wood weevil *Euophryum confine,* webbing clothes moth *Tineola bisselliella,* silverfish *Lepisma* and booklice *Liposcelis.*

The results of the trap monitoring have been used to pinpoint trouble spots and identify sources of infestation. The trap results have also been used to target cleaning and local insecticide treatments. We describe three case studies as examples to illustrate our strategies.

A CASE STUDY: THE GUERNSEY CARPET BEETLE

In September 1995, a routine check of all the traps was carried out. We were surprised to find that the traps collected from basement room B70 contained a large number of small Guernsey carpet beetle larvae, 74 in total. We increased the number of traps from two to nine so that we could identify the source of the infestation. However, over the next year all the traps caught large numbers of small larvae, some catching more than 80. The objects in the room were part of the ophthalmology collection, which are mostly composed of metal and/or wood. These were stored either on open shelving, in well-sealed cupboards or in Dexion cupboards. During the year, vulnerable objects were checked for signs of infestation but none were found.

By September 1996, we still had not found the source of the problem but the numbers of Guernsey carpet beetle larvae catches had increased to such a degree (Figure 2) that it was decided to take more drastic action. Every object was removed and checked and placed in a clean area. A few objects showed signs of active infestation. This was confined to the wool-felt material attached to the base of some of the instruments. However, one object, a

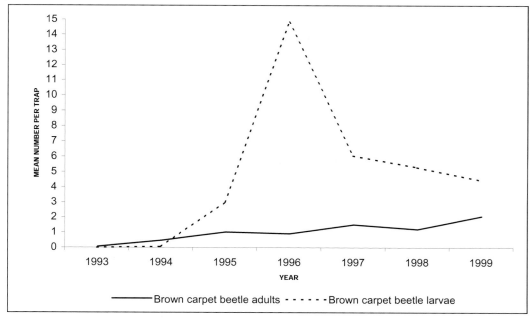

Figure 1 Mean number of brown carpet beetle adults and larvae caught in traps

tortoiseshell spectacle frame, was damaged by carpet beetle larvae. The room and cupboards were vacuumed and the floor was cleaned. We then called in a contractor to spray the edges of the room and the base of cupboards with an insecticide, Empire 20 (micro-encapsulated chlorpyriphos). After allowing the room to air for two weeks we replaced the objects. We continued to monitor the room and a few larvae were caught on the traps. We did not want to respray the room, as it is a very small space and full of objects, so we used Drione®, a silica aerogel desiccant dust containing 1% pyrethrins. Drione® is a dry dust that kills the insect by absorbing waxes and moisture which causes the insect to dry out and desiccate, rather than killing it by toxic action. Dusting with Drione® around the skirting of the room took place in January 1997 and again in June 1997. After the last dusting, no larvae have been caught on the traps. Since the problem appeared to have been controlled, the trap numbers were reduced from nine to four after April 1999. We suspect that the infestation came from dead spaces near the rooms but we never found the source.

A CASE STUDY: THE BROWN CARPET BEETLE

A problem of carpet beetle larvae was identified in the basement, in room B49, that housed the veterinary medicine, dentistry and audiology collections. The objects in the room had been housed in this location for some years. The first six traps were laid in 1993 and when they were checked three months later, we identified Guernsey carpet beetle larvae.

In January 1995, the first brown carpet beetle larvae were found. We increased the number of traps to 12 in January 1996. Due to the increasing numbers of both species of larvae, the room was deep cleaned in October 1996. This entailed vacuuming the floor, the skirting, underneath and around wooden pallets and along the window ledges. The shelves housing objects were cleaned and the objects were checked for signs of infestation. After the deep clean the numbers of larvae found decreased, particularly of the Guernsey carpet beetle, although large numbers of brown carpet beetle larvae were still being caught.

The room was again vacuumed and treated with Drione® in January 1997 and again in May 1997. The numbers of brown carpet beetle larvae were reduced but then started to rise in January 1998 (Figure 3). Looking at the trap data sheets it was apparent that the larvae were mainly caught on the traps placed along the east wall of the room. All the objects nearby were once again checked for signs of infestation but none was found.

Experience from other Drione®-treated rooms in the store indicated that larval numbers usually drop when rooms are treated, and there is normally no further reemergence. This led us to think that perhaps the problem might lie in the building structure itself. The window covers were removed but nothing was found, and so we investigated the rooms directly above, which included the staff mess room, the warders' mess room and the equipment storeroom. Two traps were placed in the staff mess room in May 1997. They were checked after three weeks and 26 brown carpet larvae were caught on one of the traps. More traps were laid in all three rooms, but unfortunately there was a problem as they were often

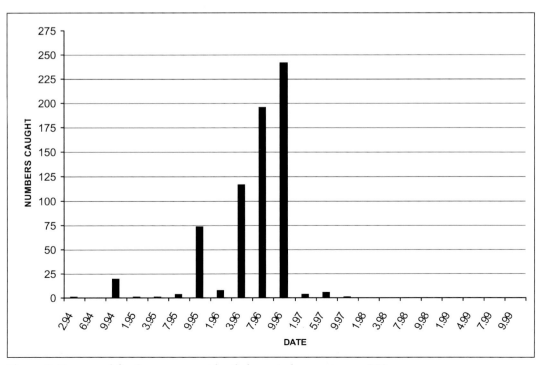

Figure 2 Trap record for Guernsey carpet beetle larvae in basement room B70

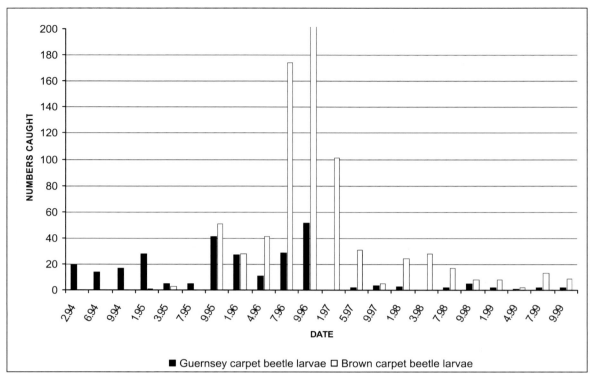

Figure 3 Number of Guernsey carpet beetle larvae and brown carpet beetle larvae caught in traps in basement room B49

removed or squashed. Brown carpet beetle larvae were the main insects to be caught in the traps (Table 1) along with a few brown carpet beetle adults and Guernsey carpet beetle larvae.

In February 1998, the staff and the warders' mess rooms were deep cleaned and treated with Empire 20, and the equipment room was also deep cleaned. In January 1999, the staff mess room was refurbished to improve hygiene and to remove dead spaces and harborages. The traps were removed and have not been replaced. Due to the continued presence of brown carpet beetle larvae in the basement storeroom and in the rooms above, the wall area was treated with Drione® in March, June and September 1998 and again in August 1999.

Although the presence of brown carpet beetle larvae has been greatly reduced, they are still present, but at least they are confined to one area. Refurbishment of the warders'

mess room and the equipment room is to be carried out in the near future, and with a new cleaning regime it is hoped that this should further reduce the problem.

A CASE STUDY: THE BISCUIT BEETLE

In 1993, we were first alerted to the possibility of an insect problem in one of the storage rooms by the number of dead biscuit beetle adults found in the basement corridor outside a series of rooms. Each room was carefully inspected and adults of both Guernsey carpet beetle and biscuit beetle were seen on the floors. The focus of the infestation was in room B35, which housed the *Materia Medica* collection. This contains particularly vulnerable material, such as dried plants and seeds. The infestation had spread to the adjoining room, which, fortunately, contained only stone and plaster objects. We placed three traps in

Table 1 Brown carpet beetle larvae trap data results

Date	Staff mess room (8 traps)	Warders' mess room (3 traps)	Equipment room (3 traps)
09/97	41	36	30
01/98	14	64	23
03/98	No traps	No traps	30
07/98	10	10	8
09/98	1	0	7
01/99	1	22	5
04/99	No traps	4	1
07/99	No traps	5	3
09/99	No traps	1	7

room B35 and, in 1993, we found adult biscuit beetles and Guernsey carpet beetle adults and larvae. We continued to monitor the room and, because of increasing numbers of insects caught, the room was deep cleaned, and every object checked and, if necessary, treated. This was a huge job as the room was packed with glass jars and paper wrappers containing the material.

In January 1994, we started transferring the contents of the jars into polyethylene bags and these were then frozen in a precooled freezer at −38°C for 72 hours. After treatment, the contents were then allowed to reach ambient temperature gradually to avoid condensation forming. Unfortunately, this took much longer than anticipated and we could only carry out the treatment when other priorities allowed. There were approximately 8000 objects to treat and this method, if continued, would have taken many man-hours. This led us to investigate a quicker method of treating the objects, bearing in mind resource restraints. Also, by 1995, the restorage of this collection had become part of a much larger project that had a fixed time to completion.

The treatment selected was fumigation by a contractor. The collection was moved to a bubble structure and exposed to an atmosphere of methyl bromide. By May 1996, all the objects had been removed from B35, the old wooden storage cases were disposed of and the room was cleaned. A contractor sprayed the whole floor area and edges of the room with Empire 20. Traps were again placed in the room and the room was resprayed in November 1996. After the room was allowed to air, new shelving was installed and a different collection, classical and medieval medicine, was brought in. The treatments controlled the infestation and the room remains pest-free (Figure 4).

After freezing or methyl bromide exposure, the *Materia Medica* objects were stored in a different room. Prior to

restorage, the room was vacuumed and dusted with Drione®. To date, we have had sporadic catches of one or two Guernsey carpet beetle larvae but no biscuit beetle adults on the traps and the collection is closely monitored.

SUCCESS OF PEST STRATEGIES: WHERE ARE WE NOW?

In June 1997, we drafted a pest strategy for both Blythe House small object store and the main site museum. The aim of the strategy was to prevent the deterioration of the collections from insect pest attack. It recommends how to integrate pest management into the museum's working practices and proposes a scheme for the control of pests. The strategy also supports the Museum's Collections Management Policy Statement. This has helped us to set out our aims and gives the relevant staff and contractors a standard to work to. As a key part of the strategy, periodic training is given to all staff involved with the collections and upkeep of the building.

Although we continued to be vigilant during 2000 there has been an increase in Guernsey carpet beetle adults although the larval catch has remained low. We have also found varied carpet beetle adults *(Anthrenus verbasci)*, as well as Guernsey carpet beetles, on the third floor in the store. The numbers of Guernsey carpet beetle adults on the fourth floor have been higher than anywhere else in the building. Checking objects has not revealed any large infestation problems. As the adult beetles are active fliers and may aggregate in the upper floors, we believe the source of the problem is in the building structure. Most of the disused ventilators and ducts have been checked, cleaned and treated, but there may still be undiscovered pest reservoirs near the roof. The number of brown carpet

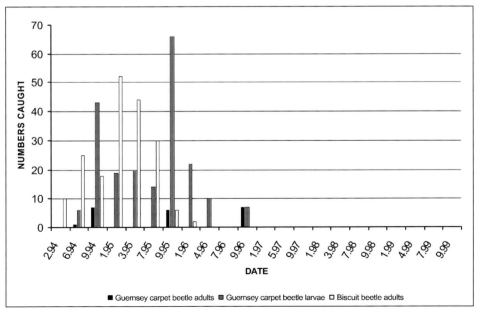

Figure 4 Number of Guernsey carpet beetle adults and larvae and biscuit beetle adults caught in traps in basement room B35

beetle adults has steadily increased, whereas the number of larvae has gradually declined since 1996.

The number of biscuit beetles has declined dramatically since the main infestation source was treated. Although the infestation was controlled in the basement, there are still occasional outbreaks of adults in other isolated areas in the building. This probably indicates that there are other minor pest reservoirs in organic debris in the building.

Since the treatment with methyl bromide in 1994, this chemical fumigant has not been used in this museum. Alternative methods for treatment of objects are freezing, fumigation with carbon dioxide and anoxia with nitrogen. A contractor carries out the carbon dioxide treatments, whilst we have developed our own on-site anoxia system using cylinders of nitrogen gas.

CONCLUSIONS

The overall picture is positive. Traps have been shown to be effective tools to locate infestation, and targeted cleaning and treatment has reduced pest numbers.

There is a proven need for the use of quarantine procedures, as movement of unchecked and untreated objects has been a clear source of pest introduction into the store. When the quarantine system has been adopted, it has clearly prevented live pests being introduced.

It is essential to understand the building and identify the effects of ducts, fireplaces and areas that are not kept clean. Often a building will attract insects, such as cluster flies and clover weevils, which are not harmful to objects but will be a food source to those that are. There is a need for widespread trapping, regular checking and constant vigilance, as new pests can appear, and others can reemerge.

The main practical recommendations are:

- set up quarantine procedures and a designated area
- maintain good housekeeping
- do not panic; assess your problem
- use insect traps and check regularly
- learn to identify common insect pests
- target cleaning
- use the most appropriate insecticide and/or gas
- train staff to raise awareness of IPM

REFERENCES

Coombes C W, Woodroffe G E, 'The effect of temperature upon the longevity, fecundity and circannual development of *Anthrenus sarnicus* Mroczkowski (Coleoptera: Dermestidae)', in *Journal of Stored Products Research*, 1983, **19**(3), 111–115.

Dvoriashina Z, 'The smirnov beetle as a pest in libraries', in *5 Restaurator*, 1988, **9**, 63–81.

Lefkovitch L P, 'A laboratory study of *Stegobium paniceum*', in *Journal of Stored Product Research*, 1967, **3**, 235–249.

Pinniger D B, Blyth V, Kingsley H, Insect trapping: the key to pest management, in *Pre-prints 3rd Nordic Symposium: Insect Pest Control in Museums*, 1998, 96–107.

MATERIALS AND EQUIPMENT

Drione®, a silica aerogel desiccant dust containing 1% pyrethrins (no longer available in the UK), Empire 20, encapsulated chlorpyriphos

> Killgerm Chemicals Ltd
> Denholme Drive
> Ossett
> West Yorkshire WF5 9NB
> United Kingdom
> Tel: +44 1924 277631
> Fax: +44 1924 264757

> Contractor: Cannon Hygiene Ltd
> Windmill Business Centre
> 2/4 Windmill Lane
> Southall UB2 4NJ
> London, United Kingdom
> Tel: +44 20 8571 0022

> Pest Traps
> Historyonics
> 17 Talbot Street
> Pontcanna
> Cardiff CF11 9BW
> Wales
> Tel: +44 29 2039 8943
> Fax: +44 29 2021 8672

ACKNOWLEDGEMENTS

Many thanks to Catherine Nightingale for her help and dedication in preparing the *Materia Medica* collection for fumigation and her valuable assistance in the treatment and deep clean of all infested areas.

BIOGRAPHIES

Helen Kingsley holds a BSc in Archaeological Conservation and Material Science from the Institute of Archaeology, London University 1984–87. Helen has worked for numerous excavations abroad and in museums in London and the United States. Helen has been at the Science Museum in London since 1992 where she is the Conservation Manager.

David Pinniger holds a BSc in Zoology from Hull University. Until 1996, David was an entomologist with the Central Science Laboratory of the Ministry of Agriculture, Fisheries and Food. He is now an independent consultant entomologist advising many major museums and historic houses in the UK on pest management strategies.

CHAPTER TEN

GREY BISCUITS, FLYING CARPETS AND CIGARETTES: AN INTEGRATED PEST MANAGEMENT PROGRAMME IN THE HERBARIUM AT KEW

Yvette Harvey

Herbarium, Royal Botanic Gardens, Kew, Richmond, Surrey TW9 3AE, United Kingdom
Tel: +44 20 8332 5238   Fax: +44 20 8332 5278   e-mail: Y.Harvey@rbgkew.org.uk
Website: http://www.kew.org

ABSTRACT

An integrated pest management (IPM) programme has been running at the herbarium, Royal Botanic Gardens in Kew, since 1997. There are 130 non-specific traps, situated in the collection areas, which are checked four times per year. Freezing is the main method of prevention, as well as for cure, in combination with a permethrin-based spray and taxon specific pheromone traps. In successive years, outbreaks of *Stagetus (Theca) pellitus*, *Anthrenus verbasci* and *Lasioderma serricorne* have been treated.

KEYWORDS

Pest management, Royal Botanic Gardens, *Lasioderma serricorne*, *Anthrenus verbasci*, *Stagetus pellitus*

INTRODUCTION

The herbarium of the Royal Botanic Gardens in Kew is an important world resource, containing an ever-expanding collection of 7.5 million pressed and dried plant specimens, which are glued to paper and filed in a systematic arrangement. The specimens, glues, papers, wooden cabinets and their door seals, all offer nourishment and/or shelter to insect pests. This presentation will 'chart' the progress of an integrated pest management (IPM) programme conducted in the herbarium since 1997.

There will be a brief introduction to the building and the collections. The insect pests commonly encountered will be discussed, along with some examples of the damage caused, and followed by a look at pest control, past and present. Data will be presented of the insects trapped during 1997–2000. Finally, there is a look at some of the annual battles waged against the insects found in the building, including grey biscuit beetles in the mycology section, flying carpet beetles eating the cabinets and cigarette beetles in the Collections Management Unit (CMU) and the Mounting Room.

THE BUILDING AND COLLECTIONS

In 1852, Hunter House, then a private dwelling built in the 18th century, had the ground floor converted for temporary accommodation of herbarium specimens and library books (Desmond, 1995). Thereafter, every 30 years or so, extensions have been added to house the ever increasing numbers of specimens in this 'temporary accommodation', so that the main building now has four large wings surrounding a courtyard (Japanese garden). The mycology collection is housed in a separate building adjacent to the herbarium. We add approximately 30,000 specimens per year (or, in terms of storage, about 60 m in height of cabinets). The whole of the building is listed and this severely restricts attempts to control the internal environment.

The exterior of the building has plantings of shrubs and climbers that provide excellent nesting sites for both birds and squirrels. The window ledges are also known to be roosting sites for pheasants and ducks! Vents around the building are unblocked for health and safety requirements. Inside, the collections are in four huge, open-plan wings (see Figure 1) and a basement. Herbarium specimens are stored in either wooden or metal cabinets in the wings and in archival boxes in the basement. Because of the vastness of the collections, for curatorial purposes the herbarium has been divided into sections, each of which curates a range of plant families. Specimens reach the sections only after being frozen in the accessions office (a separate building to the main collection) and sorted in the CMU. In the main collections area space is at a premium. Material awaiting naming, mounting and/or incorporation into the collection has historically been stored on open racking for a couple of years before being placed inside cupboards. The current practice is to store specimens awaiting attention in closed archival boxes. Bays amongst the

Figure 1 Open-plan herbarium interior (wing C). Photographer: Tudor Harwood

collections are used as working space and offices. Botanists are generally messy and cleaning is frequently hampered by clutter in the working bays.

PESTS AND THEIR DAMAGE

The main pests encountered are *Anthrenus verbasci* (varied carpet beetle), *Stegobium paniceum* (biscuit beetle) and *Lasioderma serricorne* (cigarette beetle). We also have *Attagenus pellio* (two-spot carpet beetle), *psocids* (booklice), *Trogoderma angustum* (cabinet beetle), *Anthrenus flavipes* (furniture carpet beetle) and *Lepisma saccharina* (silverfish), in addition to other opportunists. Damage to herbarium sheets includes silverfish grazing the surface of both plant specimens and labels (see Figure 2), *Anthrenus* eating through plant samples to reach insects trapped within and eating the cupboard door seals (see Figure 6), *Lasioderma*

and *Stegobium* eating the specimens and mice shredding specimens for bedding.

PEST CONTROL AT THE HERBARIUM, ROYAL BOTANIC GARDENS IN KEW

Historically, pest control involved using highly toxic chemicals. Periodically, or when an insect outbreak was detected, the herbarium was fumigated using methyl bromide. Specimens were decontaminated by 'gassing' in a small fumigation chamber just outside the herbarium, prior to entering the main collection. Originally, carbon bisulphide gas was used prior to 1937. In late 1937, following advice from Dr A B P Page of the Imperial College of Science and Technology, hydrogen cyanide, under the trade name Zyklon®, was used (Ballard, 1938). However, Zyklon® was later replaced by methyl bromide,

Figure 2 Damage to a fern specimen. Photographer: Andrew McRobb

the fumigant used until the late 1970s. Fumigation was undertaken by a rota of junior botanists. Gas masks were issued in the 1950s, but prior to this time botanists held their breath when entering the chamber! In addition to fumigation, freshly mounted herbarium sheets were painted with a solution of mercuric chloride. This continued until about 1987 when it was stopped for health and safety reasons. In the mid 1970s, Kew, in collaboration with the Central Science Laboratory in Slough, undertook pioneering work on freezing (testing the viability of insects and paper insulation) for decontaminating herbarium specimens. From the late 1970s until 1997, the herbarium decontaminated incoming material by freezing at a temperature of approximately −26°C for 48 hours.

Following advice from consultant David Pinniger and armed with the IPM 'bible' (Pinniger, 1994), the herbarium started its IPM programme in 1997. One of the first changes was made to the freezing regime. The old freezers were replaced with Vestfrost® domestic freezers that operate between −35°C and −55°C. All incoming material is now frozen for at least 72 hours at −35°C. In

addition, roughly 135 'non-specific' insect traps were placed at strategic points around the herbarium and mycology building. These traps are monitored four times per year (March, July, September/October and December/January) and are replaced annually. In 2000, *Lasioderma* pheromone traps were used in the building. Treatments are targeted at specific areas when required. The only chemical used is the permethrin-based Constrain®. IPM at the herbarium is a 'high-profile campaign' and all 200 staff are kept informed annually of the highs and lows of the trappings of the previous year.

The four-year period has seen some considerable changes to the structure of the building. The 1997 campaign went without a problem. However, the same could not be said for 1998. There was a loss of two thirds of the traps due to major electrical work throughout the building, unauthorized deep-cleaning and a changeover of cleaning contractors (and ever since, there has been no real consistency of cleaning staff). In addition to this, there was a slow leak of water from a blocked drain in one of the wings that led to condidial fungal growth over walls, cabinets and specimens in the adjacent area. In 1999, a new floor was built on top of one of the wings. And finally, the latter half of 2000 has been plagued with window repairs and painting.

TRAP RESULTS

Findings for the four years (1997–2000) can be seen in Figure 3. It can clearly be seen, in terms of numbers present, that our biggest pest has been *Anthrenus* species. The problem that we have with this beetle is decimation of cupboard door seals, as *Anthrenus* larvae eat the woollen felt that was used for the seals. The main collection storage areas and mycology section are the most affected parts of the building. *Anthrenus verbasci* is the most commonly encounted species, although a few *Anthrenus flavipes* have been trapped in the mycology collection. As can be seen in Figure 4, all areas of the buildings now have diminishing numbers of *Anthrenus*. One particular hot spot was identified in the first year of survey and its treatment is described in more detail below.

Figure 3 shows the major specimen-damaging pests. *Stegobium paniceum* has remained in low numbers throughout the four years. *Stagetus pellitus* hit a peak in 1998. This beetle was only found in the mycology

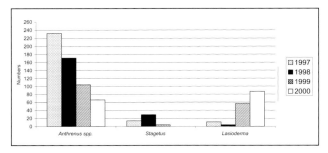

Figure 3 Numbers of insects trapped 1997–2000

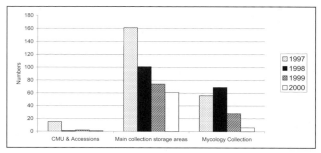

Figure 4 *Anthrenus* species trapped in specific areas

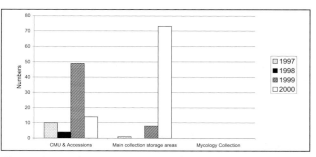

Figure 5 *Lasioderma serricorne* trapped in specific areas

collection and its elimination is described in more detail below. *Attagenus pellio* has also mostly been found in the mycology collection. Because it is only present in low numbers, no treatments have been attempted. Similarly, we have only trapped small numbers of *Ptinus* species (spider beetle). Over the last two years there have been *Trogoderma angustum* larvae present on traps in accessions and one of the wings in the main collection. A male adult was caught in the main collection in the late spring. This is the first sighting of this beetle in an English herbarium but it is a known pest in Scotland and Scandinavia.

Numbers of *Lasioderma serricorne* have been steadily increasing. Figure 5 shows the areas of the building infested with this species. As far as can be ascertained from searching the literature, this is the first known outbreak of *Lasioderma* in a major UK national institution. As can be seen, the CMU and accessions had a major outbreak in 1999, and this unfortunately appears to have led to an outbreak in the main collection area in 2000 (one wing in particular). Although the pheromone trapping has considerably boosted the number of findings, there have also been large numbers present on the non-specific traps. The beetles trapped in the CMU and accessions during 2000 were all in the accessions office. This is not a problem because the beetles should theoretically be killed in specimens when they are frozen. Dealing with the CMU outbreak in 1999 is described in more detail below.

MAJOR BATTLES WAGED 1997–2000

The 1997–8 campaign discovered a large number of *Stagetus pellitus* (formally known as *Theca pellita*) larvae and adults in the fungi collections. It was initially thought that these were a peculiar 'ecophenotype' of the biscuit beetle, *Stegobium paniceum* – superficially similar but greyer and more hairy. *Stagetus pellitus* is a Mediterranean species occurring in North Africa, southern France and Spain, etc., associated with rotten wood and associated fungi. The epicentre was the boletaceae collection. The specimens and storage boxes in and around this taxon were frozen, and the cabinets and floors thoroughly vacuumed and sprayed with Constrain® (hundreds of larvae had been discovered writhing in the compactor unit tracks). The mycology collection had previously been housed in cabinets in an air-conditioned wing in the main herbarium building. In 1995, the collections were moved to an adjacent building

where they were stored in archival quality boxes, on movable storage racking. There were no environmental controls in the new building and temperatures soared in the hot summers, especially when it was discovered that there was a problem with the boiler and the radiators were running at full-blast in June. As a direct consequence of the pest outbreak, this building now has features to reduce heat and a repaired heating system!

Over the winter of 1997–8, the hot spot of *Anthrenus verbasci* activity was tackled. Precious little of the woollen felt linings was left on the cupboards in the rosaceae and melastomataceae collections (see Figure 6). The consistent numbers of beetles trapped indicated that the food provided by the cabinets was sustaining an indigenous population of *Anthrenus*. The curators of this section painted the remaining woollen felt with Constrain® in an attempt to kill all remaining larvae and visiting adults. This has proved successful recently as there have been significantly lower numbers present in the traps in this section. In early 1999, *Anthrenus* larvae were discovered tunnelling their way into the pitchers of insectivorous *Nepenthes* specimens in order to get to the insects trapped within. Specimens from the entire family were frozen and the woollen door linings sprayed with Constrain®. It has subsequently been decided that all felt linings should be replaced with a synthetic material. It was calculated that about 20 km of material would be needed to replace all door linings in the whole building! A trial using a silicon strip is being conducted on the nepenthaceae cabinets.

In the autumn of 1998 our battle against *Lasioderma serricorne* (cigarette beetle) commenced. Infested material was discovered in an inner office of our CMU. The specimens were frozen and the office shelving was sprayed with Constrain® and all appeared to go quiet. In March 1999, *Lasioderma* was discovered in the Mounting Room. Again, all the contaminated specimens and adjacent material were frozen and the shelving sprayed, and all appeared to go quiet. Although there was no real evidence of insects on traps in the rooms, flying adults were caught in October. In addition, a number of bundles of specimens had reached the collection areas from both rooms with live beetles. All bundles were immediately recalled and frozen for a second time, and the rooms were placed in quarantine. We are very fortunate to have at our disposal a large walk-in freezer at Wakehurst Place, Kew's sister

Figure 6 *Anthrenus* larvae damage to woollen cabinet door seals. Photographer: Y B Harvey

garden in Sussex. Although devoted to the seed-bank collections, we were able to have use of the freezer during the Christmas and New Year period. We froze the entire contents of both rooms, about 1000 big boxes or two lorries full of specimens, at −40°C. Whilst the rooms were empty, they were thoroughly cleaned and sprayed with Constrain®. From the results of trappings in 2000, it unfortunately appears that some of the 1998 beetles may have escaped the bundle recall and have subsequently become established in one of the collection areas. Our next task will be to eradicate it from these areas.

CONCLUSION

Pest control at Kew has travelled a long way since the fumigation chamber of the 1930s. We now have a system that not only prevents the spread of beetles from specimens, but is also very effective at giving an early warning of outbreaks. Staff awareness of insects has been considerably raised, maintenance of the building is given priority and health and safety is taken seriously. In addition, the whole programme is very low-cost and no longer depletes the ozone layer.

REFERENCES

Ballard F, *Herbarium Specimens and Gas-poisoning,* Kew Bulletin, 1938, 397–399.

Desmond R, *Kew: The History of the Royal Botanic Gardens,* 1995, Harvill Press with Royal Botanic Gardens, Kew, London.

Pinniger D, *Insect Pests in Museums* (Third Edition), 1994, Archetype Publications Limited, London, 58.

MATERIALS AND EQUIPMENT

Non-specific 'Museum Traps' and Constrain®
  Historyonics
  17 Talbot Street
  Pontcanna
  Cardiff, CF11 9BW
  Wales
  Tel: +44 29 2039 8943
  Fax: +44 29 2021 8672

'New Serrico' pheromone lures for *Lasioderma serricorne*
  Barrettine
  St Ivel Way
  Warmley
  Bristol BS30 8TY
  United Kingdom
  Tel: +44 117 967 2222
  Fax: +44 117 961 4122
  e-mail: sales@barrettine.co.uk

Humidity-controlled heat treatment
  Thermo Lignum® UK Ltd
  Unit 19, Grand Union Centre
  West Row, Ladbroke Grove
  London W10 5AS
  United Kingdom
  Tel: +44 20 8964 3964
  Fax: +44 20 8964 2969

Low temperature domestic freezers
  Vestfrost®
  Tangley House
  Aston Rowant
  Oxfordshire OX9 5SN
  United Kingdom
  Tel: +44 1844 352906
  Fax: +44 1844 353646

A/S Vestfrost®
  Spangsberg Møllevej
  Postbox 2079
  DK-6705 Esbjerg Ø
  Denmark
  Tel: +45 79 14 22 22
  Fax: +45 79 14 23 55
  Telex: 54 122

ACKNOWLEDGEMENTS

For the past four years pest control in the herbarium has been an equally shared activity between Elizabeth Woodgyer and myself. Parmjit Bhandol joined the team in 2001. A big thank you has to go to Elizabeth and Parmjit for aiding and abetting the bug-hunts and treatments. In addition, I would like to thank colleagues at Wakehurst Place, Jonathon Farley, the CMU team and 'Steve and the lads', all of whom formed numerous chains to move approximately 1000 heavy boxes to and from the freezer to the vans, and the vans to the departments back at Kew. I would also like to thank the keeper, Simon Owens, for his continued support of our IPM programme. Martin Sands has been of immense help with the historic background to pest control.

Outside of Kew, firstly, and most importantly, I have to thank the pest guru, David Pinniger, who is always on hand to answer my endless questions. From the Natural History Museum, London, I would like to thank Phil Ackery for his invaluable assistance during the formative years of our IPM programme, Richard Adams who first identified *Stagetus pellitus* (Chevrolat) and M. Kerley and L. Rogers for the information about *Stagetus pellitus*. Finally, I would like to express my appreciation to colleagues from the UK and worldwide herbaria for the free exchange of pest information. Pest outbreaks come and go quicker than the time it takes to publish, so networks are vital.

BIOGRAPHY

Yvette Harvey has worked, as a botanist and curator, in the herbarium at the Royal Botanic Gardens in Kew since 1985. From late 1996 until the present time, accompanied by a colleague, she has been in charge of pest control in the herbarium.

CHAPTER ELEVEN

A TOPICAL SOLUTION TO TROPICAL MUSEUM PEST CONTROL

Dean Sully, Liu Man-Yee and Lee Swee Mun

University College London, Institute of Archaeology, 31–34 Gordon Square, London, WC1H 0PY, United Kingdom
Tel: +44 20 7679 7497   Fax: +44 20 7383 2572   e-mail: d.sully@ucl.ac.uk

ABSTRACT

At the National Heritage Board (NHB) of Singapore, an approach to pest control management in the museums' collections has developed alongside the relocation of existing stored collections to a new purpose-built storage building. This process of relocation provided an ideal opportunity to tackle the prevailing levels of insect activity within the museums' stored collections. The occupation of this new building has enabled staff at the Heritage Conservation Centre (HCC), to instigate an approach to integrated pest management (IPM), which has proven to be effective in establishing and maintaining low levels of insect activity.

This has evolved into a two-tiered approach. The new storage collection building is managed within a rigorously applied IPM framework. This ensures maximum control where the greatest risk occurs, whilst concurrently a more adaptive IPM policy is maintained in the museum buildings themselves. Here, the risks of pest activity have been balanced in response to the different uses of the buildings, changing curatorial use of the collections and the level of available conservation resources. Therefore, pest control and the consequent risks to the collections have had to be balanced against other museological and commercial requirements. This report will review the IPM strategy for the storage facility in relation to that applied to the other museum buildings. It will discuss the need for pest control strategies to be appropriate to the developing needs of the institution.

KEYWORDS

Environment, collections care, insects, integrated pest management (IPM), storage, tropics

INTRODUCTION

The National Heritage Board (NHB), formed in 1993, oversees Singapore's National Museums. These are the Singapore History Museum, the Singapore Art Museum, and the Asian Civilizations Museum. The collections include artefacts associated with social history, archaeology, ethnography, and fine, modern and applied art. The NHB has a stated mission to explore and present the heritage and nationhood of the people of Singapore in the context of their ancestral cultures and their links with Southeast Asia, and the world.

This is significant in terms of the care of the collections, as both indigenous and imported collections of material culture are housed in the same display and storage environments. This has necessitated a compromise when setting environmental conditions between those prevailing in the tropics and those of higher latitudes. The NHB museums are housed in 19th century buildings that have undergone a large degree of refurbishment in the last 15 years.

Between 1997 and 2000, the NHB was engaged in transferring the majority of its stored artefact collections to a purpose-built central repository, the Heritage Conservation Centre (HCC). This has provided the opportunity to confront the problem of insect infestation in the collections on a major scale for the first time. The incorporation of insect eradication and integrated pest management (IPM) strategies into the relocation project was an attempt to ensure that the new repository remains uncontaminated and the collections are not exposed to the damaging effects of insect activity.

PREVIOUS STORAGE CONDITIONS

Prior to 1997, the majority of the stored collections were held within the historic Raffles Museum building, where some rooms had been dedicated to storage of collections continuously since 1906 (Liu, 1987). The major proportion of collections has been traditionally stored in non air-conditioned rooms, ventilated by circulation fans and open windows (Figure 1).

ENVIRONMENT

Annual external climate conditions in Singapore range between 23–36°C and 65–95% relative humidity (RH)

Figure 1 Previous stores, showing open windows

(with an average of 84% RH). Non airconditioned interiors typically range between 24–26°C and 72–78% RH. These conditions would be expected to result in accelerated rates of deterioration in a range of materials, from organic to metal artefacts (Kwa, 1989), but this is not found to be the case in NHB collections.

One advantage of the environment of the humid tropics, which works in favour of the preservation of the collections, is the lack of significant annual variation. No variation, equivalent to the cyclical seasonal change experienced in temperate regions, is evident in Singapore. The relative climatic stability, at high levels of relative humidity, has been thought to be a factor in the unexpectedly satisfactory levels of preservation for some of the collections (Loo-Lim, 1991).

At the high RH levels experienced by the stored collections, mould growth could also be expected to be a problem. However, this has not been a significant risk factor, probably due to air movement and ventilation in the stores, which is maintained by ceiling fans. Air movement affects mould growth indirectly by dissipating local microclimates that would otherwise develop around substrate surfaces (Scott, 1996). The increased ventilation used in the control of mould increases the risks of exposure to polluted air, and also perhaps most importantly, pest activity. The relative effects of these differing threats need

to be considered when evaluating a strategy for collection care in the tropics.

PREVIOUS PEST ACTIVITY

Insect infestation is considered one of the most significant risks to museum collections housed in the humid tropics (Englehardt, 1990). According to NHB staff and information from past surveys, insect infestation has been endemic and has resulted in extensive damage to some areas of the collections. A sample survey of the whole collection in 1995 calculated the level of infestation in the organic object collections to be >17%. Some collections, such as the woven fibre collection, was estimated to be 50% infested, whilst the textiles collection was estimated to be 25% infested (Koestler, 1995).

The species of insects identified in NHB collections appear to be largely similar to those found in temperate museums; indeed many household or museum pests are considered to be of tropical origin (Florian, 1997). In the tropics, due to the favourable conditions for insect growth, the life cycle tends to be rapid.

The use of insect-resistant tropical hardwoods in artefacts and storage furniture, such as teak, is thought to have restricted the proliferation of woodboring beetles in some areas of the collection. However, the effects of pests can be seen quite dramatically in the condition of the fibre and basketry collections, where an infestation of cigarette beetle, *Lasioderma serricorne,* had become well established. The NHB collections, like many other Southeast Asian collections, consist predominantly of cellulose rich artefacts, e.g. textiles, basketry, and woodcarving. This provides a large potential food source for the resident insect populations (Lim *et al.,* 1990).

The insects that have been identified as a threat to the collections are listed in Appendix 1. This list is continuously updated. In addition to these pest insects, over 26 species of non-pest insects have so far been identified. Their presence indicates levels of access to the storage areas. They also provide a potential source of nutrition for pest species, such as carpet beetle and moth.

PREVIOUS PEST CONTROL

Documented cases of residual insecticide treatments for tackling pest infestations in the collections are known. For example, Lindane (gamma HCH) and Mystox (pentachlorophenyl laurate) have been applied to the surfaces of infested objects in the past. Other methods of prevention such as the use of mothballs (Naphthalene), formaldehyde, chloroform, and more recently Xylamon (Lindane) have been used as a vapour inside storage cabinets.

Between 1995 and 1997, a system of anoxic treatment using argon was in operation. In this system individual or small groups of objects are bagged and sealed in an argon-enriched atmosphere and maintained by the use of oxygen

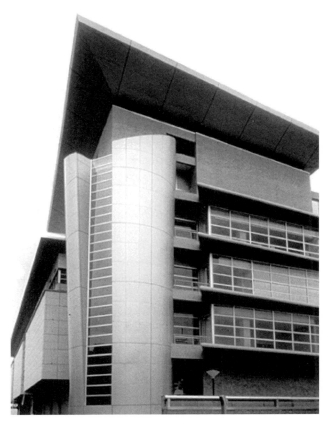

Figure 2 The Heritage Conservation Centre

scavengers below an oxygen concentration of 700 ppm for 28 days. The use of argon was recommended, in preference to nitrogen-based anoxic treatments, due to its comparatively shorter length of treatment time. In addition, nitrogen was thought to have the ability to support anaerobic growth in the high humidity situations found in the tropics, whereas argon does not (Koestler, 1995). With the adoption of this method, the eradication of active infestations in individual artefacts was tackled. However, the lack of space in the store meant that treatments had to be carried out *in situ* and the proximity to other infested artefacts meant that general levels of infestation continued. The relocation of collections to a new building (Figure 2) provided the opportunity for a larger scale solution.

PEST ERADICATION DURING THE RELOCATION OF THE COLLECTIONS

As part of the IPM strategy for HCC, an insect eradication programme was carried out to treat all vulnerable artefacts before entering the new building. Rentokil Singapore Ltd. was engaged to provide a large-scale pest eradication treatment for the relocation project. After considering the particular needs of our collections and the resources available for the project, both controlled atmosphere technology (anoxia) pest treatment and methyl bromide fumigation were chosen for use. An estimated capacity of 1600 m^3 was required to treat the selected materials.

The use of inert gas anoxic environments has been well established as a safe method for pest eradication in museum artefacts (Gilberg, 1989; Maekawa and Elert, 1996). Rentokil has developed large-scale controlled atmosphere technology (CAT) treatments using nitrogen in anoxic enclosures in Europe and the USA (Smith, 1995). The treatment of the collections during the relocation project represented the first use of this new technology in Southeast Asia. It was also the largest scale nitrogen anoxia treatment undertaken at that time. Over 40,000 objects were treated in a total volume of 500 m^3 (Smith, 1998).

Large areas, available prior to occupation in the empty storage building, provided enough space to set up these large-scale CAT bubbles (Figure 3). Each bubble was on average 35 m^3 in volume. Oxygen levels inside the bubbles were targeted to below 0.2% (0.4% oxygen being the upper level considered acceptable during treatment) for a cycle of 35 days, with a temperature of 23°C and RH of 65% inside the bubbles.

INTEGRATED PEST MANAGEMENT AT HCC

The first phase of IPM implementation has been established at HCC since 1997. It is currently being adapted to the requirements of the NHB museum buildings (Singapore History Museum, Singapore Art Museum, and Asian Civilization Museum Wings 1 and 2).

Figure 3 *In situ* CAT bubbles

IPM at HCC consists of a Buildings Pest Control Maintenance Contract and an IPM strategy for pest control for the collections of NHB.

Pest control using IPM aims to provide practical, safe and cost-effective methods that prevent collections and buildings from becoming damaged by pest infestation. The IPM strategy at HCC is based on minimizing potential pest infestation by managing the following factors.

Access: to prevent the introduction of pest infestation into the collections

All artefacts, materials and supplies that enter HCC are checked for signs of potential insect infestation before passing beyond the designated quarantine areas (see Appendix 2). This includes artefacts which are not themselves susceptible to infestation but may also provide harbourage for pests. As part of this initial check, dust or debris, especially any material from past insect infestations is removed before artefacts are moved into the permanent stores. All incoming artefacts that cannot be readily examined, or are suspected of being infested, are isolated and monitored prior to a potential eradication treatment.

Building precautions include well-sealed external envelopes with seals on their exteriors and internal doors. The building is designed without windows in all stores. Therefore, pheromone traps can provide a valuable way to identify insect populations within the storage environment.

Housekeeping: to maintain an environment that does not encourage pest infestation

The routine cleaning and removal of rubbish, e.g. from corridors, stairwells, outside areas, forms part of the building cleaning contract and is regularly monitored by the estate management team.

Food is allowed only in designated areas away from artefact holding areas and these areas are routinely cleaned. No artefacts are stored permanently on the floor or against walls, as space around artefacts is needed in order to carry out routine inspection. Flooring in the storage areas consists of white epoxy floors and light coloured vinyl flooring. The use of carpets has been restricted to office areas only.

The environment inside the storage areas at HCC is set at a temperature of 23°C with the RH at either 55% or 65%. When compared to the more favourable warm moist conditions outside, there is a considerable disincentive for insects to enter the building. This is unlike temperate museum buildings, where interior environments are generally more attractive to insect activity than outside environments.

The maintenance of relatively low internal temperature (23°C) reduces the ability for insects to fly and therefore restricts their tendency to spread. This has an implication for the positioning of pheromone traps (they need to be near floor level) and for checking for insect presence. Routine checking inside storage cabinets and boxes is required in order to identify *in situ* insect activity that might not be evident in the trapping survey.

Common corridors and outside perimeter zones are generally clean and free of dirt and debris. There is a perimeter barrier of concrete and paving and this is only breached by a small plant bed at the front of the building. This has been identified as an area that must be maintained in order to prevent a source of insect infestation developing (Pinniger, 1999).

Monitoring: to assess the extent of pest infestation and the degree of risk to the collections

Regular routine checks of the storage areas are conducted by HCC staff and potential problems are recorded. The collections most at risk, e.g. textiles and garments, organics, furniture, mixed media and paper objects, are especially targeted. The artefacts have been identified and marked to ensure that these can be more frequently inspected. Boxes and cupboards containing high-risk artefacts are regularly opened and examined.

SURVEY

As part of the monitoring process, a programme of insect trap pest surveys and sighting records is used to identify the level of insect pest activity within the collections. The extent of this survey is limited by the availability of staff time and resources. Specific pheromone traps are used to target known pest species in vulnerable stores. The checking of the traps is carried out on a three-monthly basis, and this is continuing to be reviewed in line with the perceived level of risk to the collections.

In addition, a survey of insect populations outside the building has been initiated to determine the level of pressure on the building. It can also be used to investigate the relationship between the outdoor and indoor populations. For the outdoor survey, standard and pheromone sticky traps are used.

One limitation of the trapping system is that recording, documentation, and surveying can become routinely focused on the outcome of trapping incidence. This means that the regular examination of dead areas, enclosed object storage environments, window ledges and light fittings, can become less frequently carried out when extensive trapping surveys already demand a significant investment of staff time and resources. Therefore, a more targeted location of traps in relation to the risk of the collections and a greater focus on regular general checking of the storage or display areas would perhaps provide a better level of monitoring.

How extensive the trapping survey needs to be is balanced between the risk to collections and available staff resources and how this can be integrated into general care of collections. If general housekeeping and monitoring is carried out, this information can be recorded with other relevant information in the pest control incident log. If general checks are carried out for leaks, dirt, security, etc., then the scope of the check should include observations relating to pest control.

IDENTIFICATION AND ASSESSMENT

All the insects found in the stores are retained for identification. This is critical in assessing the level of risk to the collections. It also enables the trapping survey to be adapted to the type and extent of insect activity found. Currently, insect identification forms part of the building's pest control maintenance contract and the identification service is provided on a three-monthly basis.

A spreadsheet of all insect trap catches is completed after each survey by conservation staff. If an infestation is suspected, additional traps and more frequent monitoring may be required to locate the problem. An annual report is produced from the survey data, incident logs, and any other findings during the year.

Due to the high standards of routine maintenance, cleaning, and general housekeeping, there are low levels of identified insect activity and currently no significant infestation problems. There are occasional invaders of non-pest insects and geckos, which show that there are breaches in the external building envelope that need to be identified and rectified. Monitoring, in order to identify the need for action, is the key to checking the success of this. Current high standards can only be maintained if potential future trouble spots, identified during the surveys, are eliminated swiftly and if adequate resources are provided for monitoring, maintenance, and cleaning. The present commitment by staff seen during the IPM course is a very positive factor (Pinniger, 1999).

RESULTS OF THE TRAPPING SURVEY

Since the occupation of the new building in 1997, the initial resident populations of psocids, geckos, silverfish, and cockroaches have gradually reduced in occurrence. The perennial presence of insectivores (geckos and spiders) does suggest the presence of an insect population sufficient to sustain them. These animals also tend to compete for insects with the trapping survey. Geckos are frequent visitors to the sticky traps and therefore are likely to lead to a reduction in the numbers of insects recorded in the trapping survey.

The presence of *Lasioderma serricorne* in the collections has been used in the relocation project as an indicator insect to determine the efficiency of the eradication process. The presence of this insect has decreased in the collections and the number of locations where it is present has decreased significantly. Two non-storage areas of persisting activity have been associated with inadequate external sealing of the building (roller shutters, ventilation ducting and fire exits) and also the presence of pheromone traps, which have attracted insects to enter from outside (Figure 4). Traps outside the building caught 48 *Lasioderma* in June 2000, indicating there were more pests outside the building than inside. The information from the trapping survey has been used to assess the effectiveness of current IPM strategies and to review the scope of the pest building

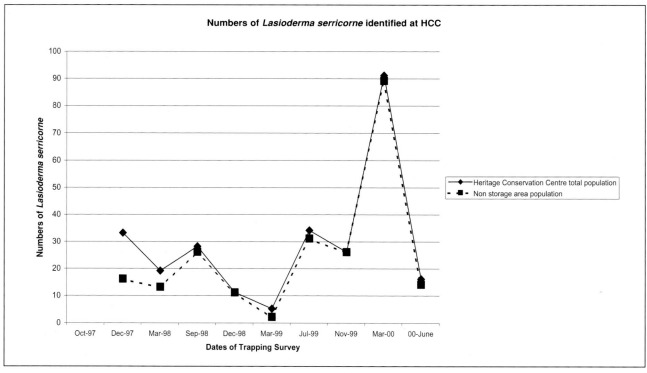

Figure 4 Presence of *Lasioderma serricorne* at HCC

management contract. This has lead to an increased sealing of internal and external access points. The survey results have reassured the NHB that the resources allocated to the relocation and eradication project were well used, and have resulted in significant benefit to the condition of the collections.

ERADICATION

Artefacts currently suspected of having pest infestation are sealed in polythene bags or sheeting, and monitored. If insect activity is identified, then the surrounding areas of the store are checked and cleaned to identify the source or cause of the infestation. In the majority of cases, the routine eradication treatment for infested artefacts is by freezing or nitrogen anoxic bagging treatment.

Freezing

Infested objects, with absorbent materials and a thermocouple, are enclosed in close-fitting, well-sealed polythene bags and placed inside the freezer. This is exposed to low temperatures at −25°C for 7 days (−30°C for 72 hours). This is intended as a treatment that ensures no survival of infesting insects. Once the use of this treatment has become more routine, the procedure can be reviewed, possibly reducing treatment times and increasing temperature, e.g. −20°C for 72 hours. Lethal cold temperature for most tropical species is suggested to be around −15°C for 48 hours (Florian, 1997).

The use of reduced temperature treatments for insect eradication in tropical environments has not been fully evaluated in a museum context. The raised moisture content of artefact materials and the fact that indigenous artefacts have not been conditioned or designed to tolerate reduced temperatures need to be considered. The danger of ice-crystal formation, when reducing the temperatures of materials with high moisture contents, and the inability of tropical insects to become cold-hardened, suggests investigating the viability of increased treatment temperatures for eradication.

Chen *et al.* (1990), examining temperate and tropical flies, identified that the tropical flies were unable to produce polyols (a type of glycerol) during lowering temperatures and therefore could not become cold-hardened (Florian, 1997).

NITROGEN ANOXIC TREATMENT

During nitrogen anoxic treatment, the infested object and oxygen absorbers are placed in an oxygen-impermeable bag. The oxygen-rich air inside the bag is removed by vacuum and replaced with nitrogen gas from a cylinder. A repeated process of sucking out air and flushing with humidified nitrogen gas continues until the oxygen level inside the bag falls below 700 ppm.

Once the required oxygen level has been reached the object is left for 30 days. The oxygen level in the bag is monitored throughout the process to ensure that the level preferably stays below 700 ppm and does not rise above a threshold level of 2000 ppm.

Once the treatment has been successfully completed, the treated object is removed from the bag, examined for insect activity, and cleaned using vacuum cleaners with High Efficiency Particulate Accumulation (HEPA) filters, before being returned to the store. Once cleaning has been completed, the treatment information is recorded on an attached artefact label. This information is then transferred to a conservation treatment record card when the object next requires interventive conservation treatment.

Our anoxic system has many built-in safeguards and can be considered as overkill. However, we hope to develop the system, adapting the timing and levels of target oxygen in relation to the type of insects undergoing treatment. The rigorous standards enforced during the introduction of a treatment process can be gradually amended once the system has proved to be effective and operating staff are comfortable with the tolerances of the system.

TRAINING

To maintain an effective pest control system, it is critical to raise the awareness of museum staff to the dangers of pest infestation and ensure that key staff are trained to perform the required functions of IPM.

As part of this process, an IPM course was conducted in November 1999 at HCC by David Pinniger and was attended by all collections care staff and key staff from other areas of the museum. In addition to the training function, the aim of engaging a pest control specialist was to review the pest control requirements of NHB museums and storage facilities. He was also able to comment on the pest control strategy developed for NHB.

IPM AT THE HCC

It has been possible to implement a rigorous IPM framework at HCC due to a number of favourable factors. These include the provision of a new, clean, and purpose-built storage facility and the chance to start 'afresh'. This raised the priority of pest control within collection care and enabled the allocation of sufficient resources to resolve the perceived problems. However, the situation in the three pre-existing National Museum buildings has been somewhat different, and policies have had to be applied to the varying needs of the individual buildings. Thus a more limited range of IPM measures are in practice for the museum buildings focused on protecting exhibition and storage areas. This is largely because the same level of access control at HCC is not currently possible at the NHB museum buildings.

GENERAL IPM APPROACH IN THE MUSEUM BUILDINGS

A building pest control maintenance contract is employed in each museum building. In addition, a pest control firm is engaged to prevent pest activity in the museum grounds. The museum precincts are fogged once a month to prevent mosquitoes from breeding. Routine inspection, baiting and spraying is conducted to control termites, flies, rodents, mice, cockroaches, geckos, birds and ants.

Pest control in the museum building storage areas is organized in a similar way to the stores at HCC. This involves the checking of artefacts before access and maintaining a quarterly trapping survey. Within the galleries, the use of trapping surveys is currently being developed to build up a picture of background insect populations. Until recently, only galleries presenting an increased risk of insect activity, during specific exhibitions, have been monitored.

IPM guidelines for the museum buildings have been developed as the basis for an initial approach to managing pest control projects. The details tend to be negotiated on a case-by-case basis. The features include the following:

- All artefacts and exhibition materials are inspected for potential insect infestation before entrance into the museum. This requires prior notice to enable checking to be carried out.
- Artefacts will only be treated if suspected or identified as being infested.
- All organic material intended for use as props or exhibition material should be checked and may be required to be treated before entering the museum.
- Vulnerable artefact material held in museum stores for medium-term or long periods should be identified to enable regular checking.
- During the construction of displays, contractors should be discouraged from using materials that encourage insect infestation, especially termites. The use of old, salvaged, or badly stored materials should be discouraged.
- The introduction of wooden materials into the building, during installation or construction work, should be monitored.
- Insect access to the museum building should be restricted, windows and doors should not be left open and poorly sealed areas should be sealed. Badly fitted or louvered or perforated windows and doors should be screened or sealed.
- Food should not be displayed, prepared, or consumed in the galleries. Events where food is served should be limited to the foyer and outside areas. All materials should be removed after the event and the area cleaned.
- Potted plants and fresh or dried flowers should not be housed in the museum buildings.

ASIAN CIVILIZATIONS MUSEUM

The Asian Civilizations Museum (ACM), opened in 1997, is the most recently established of the three national museums. It is housed in an historic school building built in the late 19th century and displays material culture from Asia.

The galleries are all air-conditioned, but only a proportion of the galleries has 24-hour climate control. Basement storerooms are conditioned with a target temperature of 22°C and 55% RH. The relative humidity of the stores and some climate-controlled galleries in ACM range between approximately 50% and 60% RH with a temperature range of 21–23°C. The stored objects are of mixed materials and are kept on open shelves.

This is a small and relatively well-sealed building, which helps to restrict the potential problems of pest entry. Another advantage of this museum is the level of awareness of pest control amongst museum staff. This has been emphasized, in part, due to the large number of high profile loans which have been used for display and which have demanded a high level of collections care and awareness of pest issues. Thus, the entry and condition checking of all artefacts entering the museum has become a routine activity.

Museum staff have been keen to safeguard lenders' collections, which often enter from non-tropical environments, against the relatively high-risk insect populations of Singapore. Thus the benefits of IPM in the ACM have been generally well received amongst museum staff.

Pest problems at the ACM

There has been a history of subterranean termite problems in this building, which had previously resulted in severe structural damage to a newly-laid oak floor. This eventually had to be completely replaced with a more insect-resistant tropical hardwood. Roof timbers had also been affected during recent years.

The particular use of the museum has also led to certain compromises in the implementation of IPM. The museum is one of the first of its kind in the region and its profile has demanded that regular public functions take place on the premises. These functions generally include catering, which has tended to be restricted to the lobby areas, and strict hygiene and cleaning has been incorporated as part of the process of each function.

Exhibition case study at the ACM

The museum highlights the various festivals celebrated by the different communities in Singapore. One of these is Deepavali, the Indian festival of light. During the festival, decorative floor designs are created using combinations of fresh flowers, spices, grains, pulses, etc., which last several weeks and thereby require a revised IPM approach. Normally, the entry and long-term presence of food materials in the museums is actively discouraged. Pest control guidelines recommend treatment of fresh vegetable materials before entry. However, due to the lack of time available it was agreed that the artists would freeze the materials themselves before use. The delicate nature of the fresh flowers meant that freezing could not be used, therefore they were left untreated. In this case, since the colourful floor patterns are an integral part of the festival highlighted in the exhibition, it was decided to monitor the gallery space using insect trapping. The designs were installed on the gallery floors with the gaps in the

floorboards sealed with a Melinex barrier to prevent ingress of food materials. The areas were thoroughly cleaned following the exhibition. No increase in insect activity was observed in the trapping survey during or following the exhibition.

SINGAPORE ART MUSEUM

The Singapore Art Museum (SAM) was opened in 1995 and highlights artists and their work from around the region, as well as hosting international travelling exhibitions. Like the ACM, it is housed in an historic building, which has been converted and adapted to the requirements of a museum.

All of the galleries are air-conditioned to 23°C and 60% RH. The RH throughout the stores and galleries range between 55%–65% RH and the temperature ranges between 22°C–24°C. Well-maintained exhibition and storage areas, with a high standard of housekeeping, help to make this museum relatively free of pest activity. IPM issues have been implemented to restrict the entry of artefacts and materials and the strict use of designated areas for public functions, which often involve the consumption of food and drinks. Like the ACM, general museum staff awareness of pest issues has been maintained since the museum was first opened. This has been further encouraged by the high proportion of loaned artwork exhibited in the museum and the associated high standards of collections care.

Pest problems at the SAM

Many of the difficulties of the museum in terms of the implementation of IPM strategies have been due to the unusual nature of many art installation works. These often involve the artist's use of ephemeral, organic materials. Examples of these, which have been exhibited in previous exhibitions, are recycled garbage, bread, rice, pastry, seeds and desiccated coconut. Such materials make an excellent potential foodstuff for insects.

In order to deal with artworks that are potential risks to infestation, a tailored IPM approach has been adopted to suit the needs of each exhibition. Trapping and monitoring is carried out in those galleries identified as at risk. However, in order to be well prepared for any potentially problematic material incorporated in an art installation, a procedure of information gathering from the artist has been in place for the previous two years. Where appropriate, each artist is required to submit a detailed description of his or her artwork including installation process, material components, and replaceable parts. This is required for loaned works as well as proposed acquisitions. Armed with such information, conservation concerns can then be negotiated whilst maintaining the artists' intent for the artwork. For example, in previous cases, the more ephemeral components of an artwork have been negotiated as replaceable, thus avoiding the need for long-term storage of materials, which are extremely vulnerable to insect attack.

Exhibition case study at the SAM

An international contemporary art exhibition of loaned material was planned for one month in February 2000. One installation art piece involved a heap of over 200 fresh red roses, information about which was not received until very late into the exhibition process. The potential problems of using fresh vegetable matter in this context are evident. Over the course of the exhibition period, one would expect the roses to decay and provide harbourage for insects. The presence of spider mites and mould growth on the roses was identified after only a few days of installation. Once conservation staff were alerted, a process of negotiation with the artist, curators and exhibition organizers began trying to avoid an escalation of insect activity. The solution was to replace the fresh roses by dried red roses that were first frozen by the Conservation department before entering the galleries.

Although this had been suggested as the preferred approach prior to the original installation, it was perhaps more useful for museum staff to see the rapid increase in insect activity occurring as a consequence of introducing fresh flowers into the galleries. The actual effect of the infestation on the insect populations in the surrounding display areas appeared minimal. The spider mites suited to pollinating fresh roses are not well suited to finding alternative niches for survival within the museum building.

The main consequence was economic, in terms of staff resources dealing with the problem and the cost of replacement dried roses, which in the tropics tend to be valued as imported exotic flowers. The economic argument of lost resources can be a powerful factor in effecting change in procedures. It is equally justified therefore, to critically review the costs involved in changing pest control procedures for the better. The consequence of a more rigid access policy, for example, is likely to have implications for planning public programmes, exhibition planning and longer lead-in times for exhibitions. This is especially significant when considering the costs of maintaining IPM programmes, which tend to be substantial in terms of staff resources.

SINGAPORE HISTORY MUSEUM

The Singapore History Museum (SHM) is the landmark museum of Singapore, having been established by Sir Stamford Raffles over 100 years ago. Over the years, gallery spaces have been adapted and refurbished, but a full-scale redevelopment of the building is currently being planned as a future project.

SHM presents the history of Singapore through the exhibition of social historical material. Most of the galleries are air conditioned to 22°C and 60% RH during public opening hours. However, the main entrance and atrium are not air-conditioned and frequently reach 28°C and 70% RH. On the exterior building walls, louvered window shutters are kept open during the day, which provide access points for insects. Old storerooms at the rear of the building are not climate controlled, and are used for the storage of props, equipment, showcases and sometimes artefacts that have yet to be processed into the museum's collections.

Pest problems at the SHM

The SHM presents problems of access and a lack of controlled environment associated with an old building in need of refurbishment. Of the three museums, it is perhaps the most challenging for implementation of best practice IPM. The entry of insects through open windows and doorways is difficult to control. In addition, the gallery spaces are only partially sealed and controlled, and therefore are potentially vulnerable. The use of gallery spaces, often 'leased' out to other organizations who install their own exhibitions, means that conservation concerns are sometimes more difficult to apply.

Exhibition case study at the SHM

In 1998, an exhibition to mark the 111th anniversary of the museum was installed. The exhibition comprised of natural history specimens on loan from another institution. Problems associated with the material (bird, butterfly, mammal and fish specimens), not only included their susceptibility to insect attack, but also the stipulations of the loan. The preferred approach to exhibiting this material was to check the condition of objects prior to installation, monitor the gallery and case interiors with insect traps and construct well-sealed showcases to restrict insect access.

Common to many natural history collections, the materials had been stored in mothball (naphthalene) filled cabinets. It was stipulated by the loaning institution that the showcases should be designed to incorporate mothballs. Problems with the use of naphthalene in terms of health hazards and effectiveness as an insect repellent have been well recorded (Linnie, 1997). However, it was important in this case to be sensitive to the lending organization's requests.

The showcases, made by a local contractor, presented further problems. The construction specifications were that they should be well-sealed, and the wood pest-free. To comply with this, the contractor treated the case materials with creosote. This was applied on the museum premises without the knowledge of museum staff. The strong lingering residues of creosote in the gallery areas were an unforeseen problem that had to be addressed with the use of air circulation and air filtering equipment.

Trapping and monitoring of this particularly vulnerable exhibition was carried out for the duration of the exhibition. This revealed the presence of *Lasioderma serricorne* within the open display platforms. The presence of this insect in the building has long been established, and its isolated presence within the gallery was not thought to compromise the safety of the vulnerable materials inside the showcases. The source of the infestation was identified and the perimeters of the affected area were treated using a commercial permethrin spray. An increased number of *Lasioderma serricorne* pheromone traps were positioned, but no further insect activity was identified in the display areas.

IPM IN THE MUSEUM BUILDINGS

The intermittent use of insect traps to check for insect activity in vulnerable exhibitions has only recently been superseded with more systematic general insect trapping surveys in the museum buildings. The numbers and types of pests found in the areas monitored so far correlate closely to the presence of environmental control. This tends to reduce insect access due to better sealing and creates a less attractive environment for insect activity to take place.

There are fewer insect pests found in the galleries and stores of the Singapore Art Museum and the Asian Civilizations Museum, compared to the Singapore History Museum. The interiors of the first two museums were specifically designed with curatorial and conservation input, whereas the oldest museum, the Singapore History Museum, is awaiting refurbishment to bring it in line with current museum standards.

The focus of IPM in the museum buildings is largely on the factors of good housekeeping, monitoring and eradication. The control of insect access in an open public building is usually compromised, therefore concentrating on identifying insect activity and taking swift action when it occurs has been a reasonable approach. So far these have been satisfactory in managing incidents of insect activity.

An extension of monitoring pest activity in the museum buildings will enable a clearer picture of the real consequences to be evaluated. Within the limitations of staff resources, activities need to be prioritized and a realistic appraisal of consequences needs to be in place.

DISCUSSION

The two-tiered approach to the pest control of the museum buildings and the HCC storage building represents an attempt to focus resources on areas of the collections at greatest risk.

The main focus of IPM activity has been in tackling the prevailing levels of infestation in the stored collections during the relocation programme. This was followed by the initiation of new IPM procedures for the new storage building. Monitoring the effect of these procedures has been a more recent priority for IPM activities at HCC. A review of these procedures has suggested the success of this process.

The nature of the museum's displays, largely based on temporary exhibitions and storage areas, housing loaned material or collections awaiting exhibition, are considered to be at less risk to long-term damage by insect activity.

The initial approach to dealing with pest control at the museum buildings has been more reactive, dealing with identified or potential problem areas. The emphasis of function of the museum buildings has been placed on public access, public activity, and facilitation of a rapidly changing temporary exhibition schedule. Therefore, IPM strategies have been required to work around these often-competing priorities. The implementation of a clear policy at HCC has helped to justify the need for changes and facilitate the broader application of IPM procedures to the museum buildings.

There are significant resource implications when applying pest control procedures for temporary exhibitions. Therefore, this should be done with a clear idea of the level of benefit in terms of averting risk to the collections.

General hygiene levels are high in Singapore, with little debris present around buildings and public areas. Generally, building maintenance contracts with pest control companies are efficiently managed by estate management staff. Along with high standards of building integrity, this limits the potential for pest insect damage in what would otherwise be very favourable conditions. However, there are endemic populations of pests such as *Lasioderma* living outdoors and so there is constant pest pressure on all museum buildings. The short-term consequences of displaying fresh foodstuffs, fresh flowers, etc., have been minor in increasing the general levels of insect activity. A realistic approach to the consequences of these activities therefore needs to be developed in relation to a realistic evaluation the risks to the collections.

Information from monitoring surveys is critical in providing justification for adapting working practices, especially where there are additional resource implications. Rapidly changing exhibition programmes, which involve the introduction of unpredictable materials that encourage pest activity, may require additional resources to resolve the problems caused. There is less risk of infestations developing and becoming damaging to the collections if the exhibition areas are isolated from reserve collection storage areas. This separation of function enables a differentiated approach to protection of different areas based on an assessment of the relative risks.

Maintaining museum buildings as multi-use public spaces demands a more flexible approach to museum pest control. The 'no food, no cut flowers' rules are inappropriate when wedding receptions are routinely held in exhibition areas and where exhibitions contain materials themselves made of foodstuff. Lead-in times of a few weeks for temporary exhibitions, demanded by the public exhibition programme, do not allow time for necessary eradication treatments.

A greater freedom in the use of the building means a

greater input of resources in monitoring the potential problems. In HCC, trapping surveys take place every three months. In temporary exhibitions, containing vulnerable material, monthly and weekly surveys take place. This is necessary to monitor initial problems with fresh material brought into the exhibition areas in the form of construction and exhibits. This enables swift action to take place to manage the problem if an increased risk is identified.

CONCLUSION

The museum situation in Singapore is not typical of the majority of tropical Southeast Asia. The NHB is responsible for the care of both locally-derived indigenous collections and a large number of significant imported collections that are housed there. The process of importing collections means that international standards of care during storage and display need to be provided (Loo-Lim, 1991). Thus, the allocation of cultural heritage resources is shaped by factors such as international competition for long-term loaned or donated collections, and the exhibition of temporary loans from institutions and individuals worldwide. This differs from an approach where human and financial resources are targeted at the challenges posed largely by indigenous collections (Pearson, 1991).

There has been a recent expansion of investment in the National Museums in Singapore, which has been focused on providing high standards of exhibition, storage and related facilities. It is an important challenge to develop collections care strategies to meet these raised expectations. The provision of modern storage and conservation facilities at the Heritage Conservation Centre has boosted the ability of the National Heritage Board to maintain its material cultural heritage to a level comparable with international standards of collections care.

In dealing with the inherent insect infestation in our stored collections, the extensive programme of pest eradication has involved, and continues to involve, considerable investment of resources. One of the responsibilities of collection care strategies is to make best use of available resources. Our approach now is to continue to monitor and maintain the collections within an integrated approach to secure this unique advantage provided by centralizing the collections of the NHB within a single storage building.

REFERENCES

Chen C, Lee R E, Denlinger D, 'The comparison of the responses of tropical and temperate flies (Diptera: Sacophagidae) to cold and heat stress', in *Journal of Comparative Physiology*, 1990, **B160**, 541–547.

Englehardt R, 'Training workshop at the Singapore and Brunei National Museums', 1990, Unpublished Assignment Report *Conservation of Cultural Property, Regional Asia*, UNDP/RAS/85/025, UNESCO, Paris.

Florian M-L, *Heritage Eaters Insect and Fungi in Heritage Collections*, 1997, James and James (Science Publishers) Ltd., London.

Gilberg M, 'Inert atmosphere fumigation of museum objects', in *Studies in Conservation*, 1989, **35**, 128–148.

Koestler R J, 'The extent of insect infestations in the collections of the National Museum of Singapore, and recommendations for treatment', 1995, Unpublished Report, Art Care International Inc., New York.

Kwa C G, 'Country report of Singapore', in *SPAFA-ICCROM Seminar on Conservation Standards in Southeast Asia*, Bangkok, Thailand, 1989, Appendix 8, 65–66.

Lim C Q, Razak M, Ballard M W, 'Pest control for temperate vs. tropical museums: North America vs. Southeast Asia', in *Pre-prints of the 9th Triennial Meeting of the ICOM Committee for Conservation*, Grimstad K (Editor), 1990, International Council of Museums, Paris, 817–823.

Linnie M J, 'Professional notes. Conservation: an evaluation of certain chemical methods of pest control under simulated museum conditions', in *Museum Management and Curatorship*, 1997, **17**(4), 414–435.

Liu G, *One Hundred Years of the National Museum, Singapore 1887–1987*, 1987, Singapore.

Loh H N, 'A historical survey of approaches to pest management in the National Museum of Singapore', in *The 23rd International Symposium on The Conservation and Restoration of Cultural Property, 1999*, Tokyo 2001, 84–99.

Loo-Lim S, 'Country report Singapore', in *Final Report of Workshop for ASEAN Conservation Laboratories, UNDP Project RAS/85/025*, Bangkok, Thailand, 28 January–1 February 1991, Appendix 10, 109–112.

Maekawa S, Elert K, 'Large-scale disinfestation of museum objects using nitrogen anoxia', in *Pre-prints of the 11th Triennial Meeting of the ICOM Committee for Conservation*, Bridgland J (Editor), 1996, International Council of Museums, Paris, 48–53.

Pearson C, 'Issues that affect cultural property, specifically objects, in South Asia and the Pacific', in *Proceedings of a GCI Symposium, Cultural Heritage in Asia and the Pacific: Conservation and Policy*, Maclean M G H (Editor), 1991, Getty Conservation Institute, Honolulu, Hawaii, 63–76.

Pinniger D, *Unpublished Report on Implementing Insect Pest Management Prepared for the Singapore National Heritage Board*, November 1999.

Scott G, 'Mould growth in tropical environments: a discussion', in *Pre-prints of the 11th Triennial Meeting of the ICOM Committee for Conservation*, Bridgland J (Editor), 1996, International Council of Museums, Paris, 91–96.

Smith C P, 'Developments in large scale anoxic treatments with nitrogen. A case study: oil paintings Marseilles, France', in *Proceedings of Meeting of Icons Working Group*, 1995, ICOM Athens, Greece.

Smith C P, 'Rentokil initial, UK', *Personal Communication*, 6 November 1998.

MATERIALS AND EQUIPMENT

Argon anoxic system
Artcare International Inc
103 Greenbush Road
Bellans Park
Orangeburg
New York 10962
USA

Insect monitoring traps
New Serrico Trap (*Lasioderma serricorne*).
Fuji Flavor Co., Ltd.
3-5-8 Midorigaoka
Hamura-sha
Tokyo 205-8503
Japan

Controlled Atmosphere Technology (CAT)
Rentokil Singapore Ltd
488 Tanglin Halt Road
Singapore 148808

ACKNOWLEDGEMENTS

The authors thank David Pinniger, consultant entomologist and Colin Smith and Kevin Peters of Rentokil, for their technical assistance. Thanks to the National Heritage Board for their encouragement to publish and to the staff of the National Heritage Board for their assistance in the project and comments on the text.

BIOGRAPHY

Dean Sully, since graduating from the University of Wales (Aberystwyth) in 1984, has worked in conservation at Chepstow Museum, the British Museum, the Museum of London, the National Heritage Board of Singapore and is now at University College, London.

APPENDICES

Appendix 1 Insect pests identified in NHB collections 1997–2000

Insects identified in NHB collections

Latin name	Common name	Type of insect	Occurrence
Ahasverus advena	Foreign grain beetle	pest species	rare, isolated population
Anthrenus flavipes	Carpet beetle	pest species	current, low
Attagenus megatoma	Black carpet beetle	pest species	current, low
Blatta orientalis	Oriental cockroach	resident species	Isolated
Blattella spp.	–	resident species	Isolated
Blattella asahinai	Asian cockroach	–	–
Blattella germanica	German cockroach	–	–
Blattella lateralis	–	–	–
Coptotermes curvignathus	Subterranean termite	pest species	rare, outside source
Cryptotermes cyno-cephalus	Drywood termites	pest species	rare, outside source
Ephestia sp.	Warehouse moth	pest species	current, low
Lasioderma serricorne	Cigarette/Tobacco beetle	pest species	frequent (most numerous pest species)
Lepisma saccharina	Silverfish	pest species	low
Liposcelis bostrychophila	Booklice/Psocid	pest species	regular
Lyctus sp.	Powder post beetle	pest species	once
Periplaneta americana	American cockroach	resident species	rare
Pheroeca interella/uterella	Wall bagworm moth	resident species	regular
Plodia interpunctella	Indian meal moth	pest species	regular
Stegobium paniceum	Biscuit/Drugstore beetle	pest species	regular
Thermobia domestica	Firebrat	pest species	regular
Tinea sp.	Case bearing clothes moth	pest species	regular
Tineidae	Micro bagworm moth	resident species	regular
Tineola walsinghami	Plaster bagworm moth	resident species	regular
Tineola sp.	Webbing clothes moth	pest species	rare
Tribolium confusasum	Confused flour beetle	pest species	rare

Appendix 2 Flow diagram of movement of artefacts into storage areas at HCC

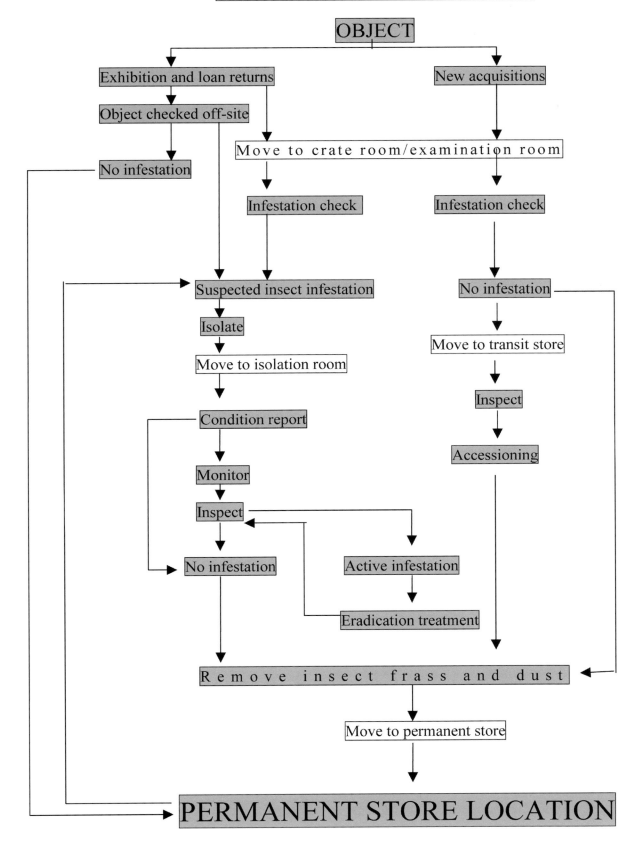

CHAPTER TWELVE

INSECT CONTROL: A TOTAL APPROACH FOR SMALL AND REMOTE MUSEUMS IN THE TROPICS

Vinod Daniel

HEAD OF THE RESEARCH CENTRE FOR MATERIALS CONSERVATION AND THE BUILT ENVIRONMENT

Australian Museum, 6 College Street, Sydney NSW 2010, Australia

Tel: +61 2 9320 6115   Fax: +61 2 9320 6070   e-mail: Vinodd@austmus.gov.au

ABSTRACT

Insects are the biggest problem for small cultural institutions in tropical countries. This paper compiles a number of practical preventive steps that can be part of an integrated pest management (IPM) plan for small museums. The paper also lists a number of non-toxic and low-toxic solutions that can be practically implemented by small museums.

KEYWORDS

Integrated pest management (IPM), insect growth regulators, low oxygen, freezing, heating

INTRODUCTION

Many small museums in the tropics preserve cultural relics of the past for future generations. The fertility of the tropical climate, however, brings its own set of conservation problems including insect pests. This, combined with large fluctuations of relative humidity, high temperatures, urban pollution, bad building maintenance and low budgets makes long-term preservation of collections a very difficult exercise for museum professionals.

Insect pests are a major problem for museums, archives and libraries, especially in the tropics. Even with repeated treatments for insect infestation, if an appropriate and safe environment is not provided for its storage or display, the insect problem will reoccur. It is therefore important for the museum personnel to devise an integrated pest management (IPM) plan for preventing the insect problems from reoccurring. It should initially be a number of simple, practical steps, which eventually can be improved upon.

This report gives a broad overview on practical methods to enable small museums to prevent insect problems from reoccurring, as well as some non-toxic and low-toxic alternatives for treating collections when a problem occurs.

INTEGRATED PEST MANAGEMENT

It is always preferable to avoid an insect problem than to deal with it once a problem occurs. An overall IPM approach will therefore be the best alternative. An IPM programme does not depend solely on pesticides to prevent or control insect problems, but instead involves the implementation of a number of measures as described by Roach (1996):

- *Physical control.* The alteration of the environment by physical means, making it hostile or inaccessible for pests. For example, insect screens and seals around doors.
- *Cultural control.* The manipulation of the pest's environment to make it less favourable. This can include relative humidity and temperature control, good housekeeping, etc.
- *Chemical control.* Appropriate selection of pesticides and their least harmful application.
- *Monitoring and evaluation* of the programme.

There are a wide range of insects that can be a problem for collections, since museums provide their basic needs, including food, water and shelter. Museum insect pests include beetles, moths, termites, cockroaches, and silverfish. The first line of defence against pest infestation must be the building itself and then the display case or storage unit. All objects brought into a museum should be carefully inspected in a quarantine (isolation) area to determine whether insects or other pests are present.

It is essential that sources of nutrition for insects, including dust and moisture, be kept to a minimum. As with the building structure, all crevices and cracks in interior fittings, around vents, ducts and piping, etc. should be sealed. Regular inspection and good housekeeping to remove dust, ensure leaking pipes are fixed quickly, etc. will help to control possible problems. It is also important that food must be kept away from working, storage and exhibition areas (Hadlington and Gerozisis, 1985; Zycherman and Schrock, 1988).

Another preventive measure is to minimize the number of trees or shrubs close to a building and to use non-flowering plant species. Gravel or paving close to the building avoids the need for watering, which in turn keeps moisture away from the building and, in addition, is non-attractive to insects and rodents. All garden rubbish and garbage from the museum must be kept well away from the building and removed or disposed of as soon as possible. Do not attach lights to a building, as any light will attract insects (Pearson, 1993).

Careful thought should be given to the layout of the building for its different functions and should make it possible to build it for insect control. Keep areas attractive to insects away from the collections (including storage and displays). The areas that attract insects include kitchens, restaurants, workshops and toilets (sources of water).

It is important to have a quarantine room where collections are held on arrival at the museum, or before treatment for insect infestation. This should be adjacent to the loading bay. In the quarantine room, the doors should be closed as much as possible. Proofing strips across the bottoms of doors would also be of benefit. Entrance doors are the other main avenues for insect entry, therefore they should be kept closed as much as possible (Pearson, 1993).

Insect-proof building materials such as brick, stone, concrete or steel should be used as much as possible. Doors should be kept closed at all times when not in use and weather or draft excluders provided, not only to keep out the weather but also to help exclude insects. Windows when closed should fit tightly, and if opened should be screened against insects. These can be excluded by the common mesh sizes of 10–20 gauge, which have apertures of 2.27–0.853 mm respectively. From the literature (Busvine, 1980; Strang, 1992a), the minimum size of holes to prevent entrance by different insect species are given in Table 1.

It is also important to remove bird nests, which can harbour insects. Other insects such as wasps, although themselves not a problem as regards the collections, leave empty nests which are a possible source of insects and so should be removed. Within the building, storage units and exhibition cases should be designed so that they close tightly. Sealing cracks and crevices of internal structures will remove breeding and hiding places for insects. It is particularly important to have good seals between public areas such as exhibition rooms, sales areas, food services, etc. and collection storage (Pearson, 1993).

Table 1 The minimum diameter of mesh hole needed to exclude insects

Mesh hole diameter (mm)	Type of insect prevented
2.3	house fly, blowfly
1.15	mosquito
0.85	sandfly
0.7–1.7	beetle (depending on the species)

INSECT TRAPS

One of the most valuable tools of an IPM programme is regular, thorough inspection. During inspections many insect problems can be discovered before too much damage has occurred. Inspections can, however, be time consuming, especially in large collections where the organic materials, and even some of the inorganic materials, are at risk from insect attack. It is here that insect traps can be of assistance.

Blunder traps are non-specific traps, which assist in catching any insects present within the collection. In spite of the presence of a food attractant in most traps, the capture is largely due to the location and placement of the traps where insects are common. The most common traps are usually made from a piece of cardboard, one or both sides of which are sticky, but many other types are available.

Trapped insects can be identified using one of the many insect identification books, or by contacting the entomology department of a museum. Through correct identification it is possible to learn whether or not the insect poses a threat to the collection, what type of material(s) are likely to be infested, and what to look for, i.e. adults or larvae.

Many traps now incorporate a pheromone as an attractant to lure insects into them. Pheromones are chemical messengers, which are produced by insects to communicate messages. These chemicals when passed from one insect to another of the same species cause a certain response, either behavioural or physiological. Some examples of these are aggregation pheromones (which may attract both males and females, for example, to a food source), trail marking pheromones (such as those used by termites and ants) and sex pheromones (which attract a mate and may initiate mating).

Any insects within a certain distance of the pheromone trap will 'home in' on the odour and will become trapped. By checking these traps on a regular basis, it is possible to get an indication of the presence (or absence) of a specific insect within the monitored area. Pheromones are now available commercially for a number of museum pests such as cigarette and drugstore beetles, common clothes moths and cockroaches.

TREATMENT OPTIONS

Chemical methods

There are a number of chemical treatment options which would be applicable for small museums. Museums can refer to many published papers on this topic, which highlight the range of chemicals from ethylene oxide to permethrins (Story, 1985; Pinniger, 1994).

Insect growth regulators (IGRs) are chemicals with low mammalian toxicity, which work specifically on insects and a few other anthropoids by disrupting their life cycles. These could include use of juvenile hormone analogues to make pupation unsuccessful or incomplete, or chitin synthesis inhibitors to prevent the formation of a normal cuticle causing death by disruption or dehydration. IGRs that are available in some countries include methoprene, hydroprene, fenoxycarb and triflumuron.

Freezing

Objects are sealed in polythene bags to ensure there is no change in moisture content and to avoid condensation on thawing. They are placed in a freezer and left for two weeks at a temperature below −20°C. This will kill all stages of the insects (eggs, larvae, pupae and adults). Once removed from the freezer, the objects are allowed to acclimatize to room temperature and then removed from their bags (Strang, 1992a and Strang, 1992b).

The effect of freezing on museum collections has been an area of many debates. There are two issues that one needs to be aware of:

- Polymeric materials may become stiffer and more brittle at low temperatures if they go through a glass transition temperature. Drying oil films have a glassy-rubbery transition between −30°C and 0°C. Acrylics used in most paint formulations are glassy below 0°C. Hence it will not be advisable to use freezing for acrylic and oil paintings. It is also important that artefacts are not placed on top of each other, and are handled with care when in a frozen condition.
- An artefact needs to be bagged to keep its moisture content constant when treated by freezing. If the material is not bagged, the moisture content may increase by about 5–6%, causing a 3% dimensional change, which is dangerously close to the material's elastic limits. (Michalski, 1992; Strang, 1995).

Freezing is a safe, non-toxic method for pest control for many collections. Here are a few categories where one needs to observed greater caution when freezing or look for alternative methods:

- paintings on canvas
- ivory
- ancient and deteriorating glass and glass components
- waterlogged specimens and artefacts
- thick powdery and/or matt paints with relatively little binder which have been painted on wood substrates
- paintings on joined wooden panels
- objects with wax components or large wax fills

Many waxy objects undergoing cycled changes in temperature can produce a polymorphism, resulting in an opaque, powdery wax formation on the surface. Wax components are often brittle and cannot withstand dimensional change if they are built up upon a wood or other organic substrate (Kronkrite, 1989).

High temperature

As with freezing, objects are sealed in polythene bags, but in this case are heated to a temperature of at least +52°C for four hours. This will kill all stages of insects. The objects are allowed to return to room temperature and then removed from the bags. It should be possible to utilize the heat from the sun in tropical countries to provide the energy for this treatment. A black plastic bag with another clear plastic bag enclosing it (Strang, 1995), or a tin shed in the museum grounds (Strang et al., 2000) may be sufficient. These containers must be fully exposed to the sun and raised off the ground to prevent heat loss. Design specifications for a solar heat treatment building include:

- Metal construction with thin materials on the roof and walls facing the sun, and insulated materials with internal heat reflectors on the other walls and the base.
- Shed painted black to promote heating effect of the sun.
- Means of measuring the temperature inside the shed – thermometer or thermocouple. Ideally the temperature in the centre of the object being treated is where the measurement is required. This is relatively easy for a collection of textiles, but will require a substitute of similar thickness and density for solid objects. Here the temperature is measured from a hole drilled into its centre.
- Means of venting and cooling the shed and contents once treatment is over. From calculations of the angles of elevation of the sun it may be possible to use nearby shade or the time of day for this purpose, for example, heat in the afternoon, and allow it to cool overnight.

The above thermal treatment methods are appropriate for single objects and collections but this is a different matter when rooms or entire buildings are infested. There are companies which can successfully heat-treat an entire building, although the process is expensive (Pinniger, 1996).

Unbagged artefacts subject to heating may undergo dimensional changes based on the change in equilibrium moisture content. Therefore, it is important that the artefacts be bagged before heat treatment. The effect of high temperature on objects is an area that is being extensively studied at this time. One obvious area where extreme care has to be taken is for materials which have a low melting point, for example, waxes.

Low oxygen systems

Considerable research has been conducted with the use of modified atmospheres to manage insect pests in stored grains and food. In experiments sponsored by The Getty Conservation Institute (GCI), Rust et al. (1996) evaluated the mortality of all life stages of ten common insect species (webbing clothes moth, furniture carpet beetle, firebrat, cabinet beetle, larder beetle, cigarette beetle, confused flour beetle, cockroach, powderpost beetle, and western drywood termite). The experiments were carried out at 55% relative humidity and 25.5°C in a nitrogen atmosphere containing less than 0.1% oxygen. The time required for 100% of the specimens to be killed varied from 3 hours for the adult firebrats to 192 hours for the eggs of the cigarette beetle. Based on these studies and other mortality data in low oxygen environments, it is recommended that two weeks is the time required for effective treatment in a low oxygen environment (less than 0.3% oxygen in nitrogen) at 55% relative humidity and 25°C.

The basic procedure for producing and maintaining a reduced oxygen atmosphere for treating museum objects is to replace air with an inert gas in the bag that encapsulates an infested object. There are three variations in protocol:

- *Dynamic system*. An inert gas is used to flush all the air out of the bag (or chamber) by an initial high flow rate and then, when a level of less than 0.1% oxygen is reached, the flow is reduced to that required to maintain the low-oxygen atmosphere for a period of the treatment (Daniel *et al.*, 1993a and 1993b).
- *Dynamic–static system*. The bag is purged with an inert gas, as in the dynamic system and once the 0.1% oxygen level is reached, a quantity of Ageless® oxygen scavenger (Daniel and Lambert, 1993; Lambert *et al.*, 1992) is quickly inserted, the gas flow stopped and the bag sealed for the required exposure period.
- *Static system*. This method is ideal for treating small objects (less than 100 L). No purging of air in the bag is necessary. A calculated amount of oxygen scavenger needed to absorb the oxygen in the bag and maintain the oxygen concentration at less than 0.1% for the fumigation period is inserted (Daniel *et al.*, 1993a and 1993b).

THE BIG PICTURE

Small museums have limitations, both in terms of financial as well as human resources. Very often they are totally managed by a single staff member or volunteer. While it is impossible to totally eradicate insect problems in small cultural institutions, the practical steps highlighted in this paper are guidelines that can minimize the damage caused by insects. It is also important to note that the damage caused by insects is usually worse than that caused by any particular treatment. Therefore, it is recommended that any affordable pest control treatment option be pursued with necessary precautions rather than debate about which treatment option is better.

In many developing countries, there are a number of traditional methods for pest control that have been pursued for generations. IPM programmes have to be adapted to use these traditional methods, as well as deal with ethical issues. Examples include the use of leaves from the neem tree in India, and the ceremonial use of tobacco smoke by the American Indians as a pest control option. In countries such as Bhutan, there are ethical issues on killing insects, which have to be managed before pest control is implemented.

It is also important when training personnel from small museums about pest control, that detailed information on the availability of local suppliers, as well as procedures for treatment, are clearly provided. This would save time for museums to research suppliers, as well as provide continuity when staff changes occur (especially with volunteers).

Overall, insects are the biggest problem for collections in small tropical museums. A cheap, practical low-toxic programme developed to minimize damage to cultural collections by insects will greatly assist in preserving the past for future generations.

REFERENCES

Busvine J R, *Insects and Hygiene. The Biology and Control of Insect Pests of Medical and Domestic Importance,* 3rd edition, 1980, Chapman and Hall, London.

Daniel V, Lambert F L, 'Ageless oxygen scavenger: practical applications', in *WAAC Newsletter,* May 1993, **15**(2), 12–14.

Daniel V, Hanlon G, Maekawa S, 'Non-toxic fumigation of large objects', in *21st Annual Meeting of the American Institute of Conservation,* 1993a, 31 May–6 June, Denver, Colorado.

Daniel V, Hanlon G, Maekawa S, Preusser F, 'Nitrogen fumigation: a viable alternative', in *International Council of Museums, 10th Triennial Meeting,* 1993b, 22–27 August, Washington DC, USA.

Florian M L, 'The freezing process – effect on insects and artefact materials', in *Leather Conservation News,* 1986, **3**(1), 1–17.

Hadlington P W, Gerozisis J, *Urban Pest Control in Australia,* 1st edition, 1985, New South Wales University Press, Sydney.

Kronkrite D, 'Museum artefacts and the deep freeze', in *Bishop Museum Conservation Newsletter,* Spring 1989.

Lambert F L, Daniel V, Preusser F D, 'The rate of absorption of oxygen by Ageless; the utility of an oxygen scavenger in sealed cases', in *Studies in Conservation,* 1992, **37**, 267–274.

Michalski S (Ed.) *A Systematic Approach to the Conservation (Care) of Museum Collections, with Technical Appendices,* 1992, Canadian Conservation Institute, Ottawa, Canada.

Pearson C, 'Building out pests', in *AICCM Bulletin,* 1993, **19**(1/2), 41–55.

Pinniger D, *Insect Pests in Museums,* 3rd edition, 1994, Archetype Press, London.

Pinniger D, 'Insect control with the Thermo Lignum® Treatment', in *Conservation News,* 1996, **59**, 27–29.

Roach A, 'Report for the training course "Pest Control in Museums" ', April 1996, Australian Museum.

Rust M K, Kennedy J M, Daniel V, Druzik J R, Preusser F D, 'The feasibility of using modified atmospheres to control insect pests in museums', in *Restaurator,* 1996, **17**(1), 43–60.

Story K O, *Approaches to Pest Management in Museums,* 1985, Smithsonian Institution, Washington DC, USA.

Strang T J K, 'Museum pest management', in *A Systematic Approach to the Conservation (Care) of Museum Collections,* Michalski S (Ed.), 1992a, Canadian Conservation Institute, Ottawa, Canada, 1–27.

Strang T J K, 'A review of published temperatures for the control of insect pests in museums', in *Collection Forum,* 1992b, **8**(2), 41–67.

Strang T J K, 'The effect of thermal methods of pest control on museum collections', in *Pre-prints of the 3rd International Conference on Biodeterioration of Cultural Property,* Bangkok, July 1995, 199–219.

Strang T J K, Mitchell J, Pearce A, Pearson C, 'Low cost methods for insect pest control', Poster at IIC Congress *Tradition and Innovation: Advances in Conservation,* 10–14 October 2000, Melbourne, Australia.

Zycherman L A, Schrock J R (Eds), *A Guide to Museum Pest Control,* Foundation of the American Institute for the Conservation of Historic and Artistic Works and the Association of Systematic Collections, 1988, Washington DC, USA.

BIOGRAPHY

Vinod Daniel is the Head of the Research Centre for Materials Conservation and the Built Environment at the Australian Museum, Sydney. He is also the Deputy Chairman for Ausheritage (Australia's cultural heritage network), council member for the Australian Institute for Conservation of Cultural Materials (AICCM), board member of the International Council for Biodeterioration of Cultural Property, and editor of the AICCM Journal. He has travelled extensively in the Asia–Pacific on various preventive conservation initiatives. His particular areas of interest are in pest control, environmental standards and monitoring. He has published extensively on these topics.

CHAPTER THIRTEEN

PRACTICAL METHODS OF LOW OXYGEN ATMOSPHERE AND CARBON DIOXIDE TREATMENTS FOR ERADICATION OF INSECT PESTS IN JAPAN

Rika Kigawa, Yoshiko Miyazawa, Katsuji Yamano, Sadatoshi Miura,
Tokyo National Research Institute to Cultural Properties, 13–43 Ueno-park, Taito-ku, Tokyo 110-8713, Japan
Tel: +81 338 23 4875   Fax: +81 338 22 3247   e-mail: rkigawa@tobunken.go.jp
Website: http://www.tobunken.go.jp

AND

Hideaki Nochide, Hiroshi Kimura and Bunshiro Tomita
KIKA Carbon Dioxide Co., Ltd., 1–2 Kiyoku-cho, Kuki-shi, Saitama 346-0035, Japan
Tel: +81 480 21 8805   Fax: +81 480 21 8807

ABSTRACT

This paper reports on an attempt to make practical protocols for low oxygen atmosphere and carbon dioxide treatments to eradicate insect pests in Japanese museums. Major insect pests were placed into three groups based on their mortality data. For each group we set up treatment conditions to achieve 100% mortality. Experiments were also carried out to examine treatment conditions for insects that tunnel deep inside books and wooden objects.

Experiments at ambient temperature (about 20°C), as well as at 25°C or 30°C, were carried out to determine the treatment period in museum storage areas.

An automated fumigation chamber, with an attached nitrogen generator, for the eradication of insect pests in cultural objects is assessed. Examples of oxygen scavengers, such as RP system® (RP Agent-K type) were also assessed. For carbon dioxide treatments, we have successfully used tents made of a barrier-film which has low permeability to carbon dioxide.

KEYWORDS

Pest control, anoxia, carbon dioxide, Ageless®, RP system®

INTRODUCTION

Japan has many kinds of insect pests that infest museum objects. They include some insect species that are common in tropical countries: termite, cockroach, silverfish, bookborer anobiid, pubescent anobiid, cigarette beetle, powderpost beetle, bamboo powderpost beetle, carpet beetle, clothes moth and booklouse. Furthermore, we often suffer from mould infestations because of the humid climate in summer months. For about 30 years, fumigation of whole buildings, using a mixture of methyl bromide (86 wt.%) and ethylene oxide (14 wt.%), was widely performed as a powerful and effective control measure. About 70,000 kg or more of the mixed gas was used annually in cultural institutions. However, pest-control strategies are now facing a period of great change, as the fumigant methyl bromide will be phased out by the year 2005. We are now trying to introduce integrated pest management (IPM) policies so that we can decrease the amount of the gas used, develop safer management policies in museums and implement other practical alternatives to deal with active infestation in museums.

Among various alternatives, we report here our attempts to establish practical protocols for low oxygen atmosphere and carbon dioxide treatments, which are suitable for the situations in Japan. The criteria for selecting these methods was their wide applicability to various kinds of materials and safety to humans. Low oxygen atmosphere and carbon dioxide treatments are being employed by cultural institutions in the United States, Canada and Europe (Selwitz and Maekawa, 1998). Low oxygen treatment using an inert gas, such as nitrogen, is thought to be one of the safest methods for museum materials when used with humidifying devices. When big tents are employed to treat large batches of material, carbon dioxide is the chosen gas, because gas concentrations effective for killing insects are more readily achieved and are cheaper (Selwitz and Maekawa, 1998). However, there is some concern about the formation of carbonic acid when carbon dioxide encounters water during treatment (Reichmuth, 1987). As sensitive materials could be harmed by acidity (Kigawa *et al.*, 1999a), we should be aware of this factor when planning carbon dioxide treatments.

The possible constraints of these methods are relatively high costs including personnel expenses during long treatment times and long treatment periods to achieve 100% mortality. However, these methods would be effective alternatives, especially where the scale of infestation is small and limited and long treatment is possible.

There are many references that describe the mortality data of common museum insects in the United States and Europe (Gilberg, 1989; 1990; 1991; Rust and Kennedy, 1993; Valentin, 1993; Smith and Newton, 1991). Although they greatly help us to implement the treatments, we needed to examine the time to achieve 100% mortality of common insects, specific to Japan, which infest cultural objects. The minimum time for 100% mortality varies from species to species and also for different life stages. Temperature, relative humidity (RH) and oxygen or carbon dioxide concentration also significantly influence the minimum time for 100% mortality (Bailey and Banks, 1980). Since it would be difficult for museum staff to follow all of the mortality data, we aim to make a simple protocol to facilitate treatments in cultural institutions. We also report on the use of an automated treatment chamber with an attached nitrogen generator for the eradication of insect pests in cultural objects, provide examples of

Table 1 Protocol for low oxygen and carbon dioxide treatments of common insect pests that infest Japanese cultural objects

	Species	Low oxygen★	Carbon dioxide★★
Group A	# Cigarette beetle (*Lasioderma serricorne*) # Drugstore beetle (*Stegobium paniceum*) Pubescent anobiid (*Nicobium hirtum*) Bookborer anobiid (*Gastrallus immarginatus*) # Brown powderpost beetle (*Lyctus brunneus*)	30°C for 3 weeks	25°C for 2 weeks
Group B	# Black carpet beetle (*Attagenus japonicus*) Varied carpet beetle (*Anthrenus verbasci*) American cockroach (*Periplaneta americana*)	30°C for 1 week or 25°C for 2 weeks	25°C for 1 week
Group C	# German cockroach (*Blattella germanica*) # Webbing clothes moth (*Tineola bisselliella*) # Japanese termite (*Reticulitermes speratus*) Casemaking clothes moth (*Tinea translucens*) Oriental silverfish (*Ctenolepisma villosa*) Firebrat (*Thermobia domestica*) # Flattened booklouse (*Liposcelis bostrychophilus*)	25°C for 1 week	25°C for 1 week

★ Oxygen concentration below 0.1% with Ageless® Z oxygen scavenger, or near 0.2% with nitrogen gas humidified to about 55% RH
★★ Carbon dioxide concentration at 60% volume with oxygen concentration at about 8%
The insects for which we confirmed mortality

treatment with a new type of oxygen scavenger, RP system® (RP Agent-K type) and provide examples of carbon dioxide treatments in tents.

PROPOSAL OF A PROTOCOL FOR TREATING INSECTS IN JAPAN WHERE ELEVATED TEMPERATURE CONTROL (25°C OR 30°C) IS AVAILABLE

We report here on an attempt to set up a simple protocol to assure 100% mortality where temperature control is available.

Gilberg (1990) proposed a treatment time of three weeks at 30°C with an Ageless® oxygen scavenger, which achieves an anoxic environment with an oxygen concentration below 0.1%, as a simple procedure to assure 100% mortality of common museum insects. We tested *Sitophilus zeamais*, one of the most resistant species to anoxia environments, and several other museum insects at these conditions and the results were very effective. However, in many cases, the time required for treatment is sometimes very critical for practical implementation of treatments under the pressure of tight exhibition schedules. In addition, the temperature of 30°C will not always be appropriate for treating delicate fine art objects. Therefore, we needed to investigate treatments at lower temperatures.

For carbon dioxide treatments, 60% is the concentration generally considered to be most practical for the control of stored-product insects (Smith and Newton, 1991). Susceptibility to carbon dioxide is also very different from species to species. We divided the major insect pests into three groups according to their mortality data based on previous reports (Gilberg, 1989; 1990; 1991; Rust and Kennedy, 1993; Valentin, 1993; Smith and Newton, 1991) for our experiments with a low oxygen atmosphere (maximum 0.2% of oxygen) and carbon dioxide (60% of carbon dioxide and about 8% of oxygen) treatments. For each group, we set up treatment conditions to achieve 100% mortality (Table 1). The insects for which we confirmed mortality can be seen in the table. The data of this table is partially revised from that in a previous report (Kigawa *et al.*, 1999c).

The proposed periods for low oxygen treatments are rather longer than those previously reported to ensure 100% mortality of some insects. The reasons are as follows. Firstly, the maximum oxygen concentration of our low oxygen treatments is 0.2%, which is higher than those of reported concentrations (below 0.1% for Rust and Kennedy, 1993; approximately 300 ppm for Valentin, 1993). Secondly, the wild strains of insects could be more tolerant to treatments than cultured insects and we therefore needed to set longer treatment times.

Conversely, our proposed periods for carbon dioxide treatments are shorter than those previously reported (Valentin, 1993). For example, in Valentin's paper, 10–25 days at 30–35°C was needed for a complete kill of

Coleoptera. The difference in mortality seems to be an interesting phenomenon. We think that the reason for the difference may be because of the oxygen concentration during the carbon dioxide treatment. In Valentin's paper, treatments were done in 60% carbon dioxide at 40% RH and 0.03% oxygen. Our conditions were 60% carbon dioxide at about 30% RH and about 8% oxygen. We observed that when the oxygen concentration was very low (especially <1% oxygen), it became very difficult to achieve a complete kill of various stages of *Lasioderma serricorne* by carbon dioxide treatments (Nochide *et al.*, unpublished results). Similarly, there are references mentioning that carbon dioxide is more toxic when oxygen is present against some of the difficult to control species of stored product pests (Bell, 1984; 1996; Soma *et al.*, 1995). Therefore, for effective carbon dioxide treatment of difficult to control species, some amount of oxygen concentration (at least for our experiments, about 5–8% oxygen) will be necessary.

The data in Table 1 is partially used as a treatment standard in an insect eradication manual by RP Agent-K type by Mitsubishi Gas Chemical Company, Inc., Japan (Mitsubishi Gas Chemical Company Inc., 1999).

TREATMENTS AT AMBIENT TEMPERATURE (20°C)

While low oxygen atmosphere and carbon dioxide treatments are more effective at elevated temperatures (25°C or 30°C), there are not many reports on the efficacy of the treatments at ambient temperature except for a few papers (Valentin, 1993; Newton *et al.*, 1996; Umney, 1997). Because the common temperature in our storage areas is around 20°C, for on-site treatment it is important to know the time required for a complete kill of our museum insect pests. In our previous study with *Sitophilus zeamais*, which we use as a test insect because it is one of the most tolerant to low oxygen atmosphere, a very small population (below 5% of the control population) survived even after a five-week treatment of low oxygen atmosphere of below 0.1% oxygen at 20°C (Kigawa *et al.*, 1999d). This is similar to the results reported by Umney (1997), which describe mortality data for another rather tolerant species, *Anobium punctatum*, with low oxygen treatment at 20°C. To determine the time to ensure a complete kill of the most resistant species, we again used various developmental stages of *Sitophilus zeamais* and examined mortality after longer treatments (eight or ten weeks) at 20°C. The results are shown in Table 2. After treatment for eight weeks by oxygen scavenger Ageless® Z or RP Agent-K type, over 99.8% of control population were killed. With a treatment of ten weeks, 100% mortality of treated insects was observed.

From this data, we consider the safe treatment time for a low oxygen atmosphere to be from approximately ten weeks at 20°C for the most tolerant group of our museum insects. Therefore, we recommend more than two months

Table 2 Emergence of adult *Sitophilus zeamais* after treatments at 20°C, with 60% RH

Test 1	Treatment period	Total emergence after 12 weeks
Control	4 weeks	439
	6 weeks	531
	8 weeks	575
	10 weeks	545
Ageless® Z	4 weeks	39
	6 weeks	8
	8 weeks	1
	10 weeks	0
RP Agent-K type	6 weeks	3
	10 weeks	0

Test 2	Treatment period	Total emergence after 12 weeks
Control	4 weeks	1806
	8 weeks	1665
RP Agent-K type	4 weeks	709
	8 weeks(1)	2
	8 weeks(2)	0

of treatment in a low oxygen atmosphere below 0.2% for insects of group A in Table 1. However, for more susceptible species, a much shorter time would be enough (Valentin, 1993; Umney, 1997). Based on their results and our experiments, four weeks of treatment below 0.2% oxygen at 20°C would be sufficient to assure 100% mortality of the other insects (groups B and C).

With regard to carbon dioxide, three weeks of treatment at 20°C in 60% carbon dioxide with about 8% of oxygen was required for the complete mortality of various developmental stages of cigarette beetle, *Lasioderma serricorne*, and biscuit beetle, *Stegobium paniceum* (Nochide *et al.*, unpublished results). However, a longer time might be necessary to kill some kinds of woodborer insects.

TREATMENTS OF BOOKS AND WOODEN OBJECTS

We also carried out experiments to examine treatment conditions for insects that tunnel deep inside books and wooden objects. All the stages of *Sitophilus zeamais* were put into holes inside books or wooden pieces, and mortality was checked for each treatment condition (Table 3). The results of insect mortality suggested that the oxygen concentration quickly fell inside books in a low oxygen atmosphere. Therefore, books may be treated with the usual treatment condition of low oxygen, which is consistent with Valentin's results (Valentin, 1993).

On the other hand, it seemed that in some cases much more time was required for depleting oxygen inside wooden pieces. We found that some insects survived after Ageless® Z treatment for three weeks at 30°C in coated wooden pieces with rather high density (Table 3). Therefore, density as well as size of a wooden piece, and

coating would seem to be critical factors for efficacy of treatment. Coated wooden pieces required longer time for the displacement of oxygen than uncoated pieces, which is consistent with the data previously reported (Selwitz and Maekawa, 1998). For example, oxygen concentration inside a coated *hinoki* piece (50 × 50 × 100 mm, density of 0.46 g/ml) was more than 1% after one week of Ageless® Z treatment, but that of an uncoated *hinoki* piece was about 0.4%. However, since there are usually many insect holes inside infested wooden objects, the oxygen concentration would be likely to decrease more rapidly than in the pieces we observed.

In the carbon dioxide treatments, it seemed easier for the gas to penetrate into books and wooden pieces to achieve the concentration required for mortality (above at least 40% volume). However, since it was reported that carbon dioxide treatment required much more time to ensure a 100% mortality of wood-infesting insects, including Anobiid and Lyctid boring beetles (Valentin, 1993), it would be very useful to confirm mortality in infested wooden objects, in order to suggest useful protocols.

In practice, since wooden cultural objects are very different in size and in wood type, more experiments will be necessary for confirming the conditions for 100% mortality of insects involved in wood infestation.

PRACTICAL TREATMENTS

Automated nitrogen treatment chamber with humidifying devices

We installed an automated fumigation chamber, with an attached nitrogen generator, for the eradication of insect pests in cultural objects by anoxia at the Tokyo National

Table 3 Emergence of adult *Sitophilus zeamais* after treatments

	N$_2$ (ca.800 ppm of oxygen) 30°C, 3 weeks			Ageless® Z or RP Agent K-type (< 0.1% of oxygen) 30°C, 3 weeks			Carbon dioxide (ca. 60 vol.% CO$_2$ + 8 vol.% O$_2$) 25°C, 2 weeks		
	Rice (g)	Control	Treatment	Rice (g)	Control	Treatment	Rice (g)	Control	Treatment
Control a	3.8	147	0	–			2.5	56	0[1]
	3.8	138	0				2.5	62	0
	3.8	144	0						
	3.8	129	0						
Control b	10	816	0	–			–		
	10	906	0						
Book a[2] (210 × 300 × 35 mm)	3.8	144	0	–			–		
	3.8	98	0						
Book b[2] (210 × 300 × 40 mm)	3.8	134	0	–			5.0	110	0[1]
	3.8	98	0						
Wooden piece a[3] (50 × 50 × 100 mm, *hinoki*, density 0.46g/ml)	2.0	69	0[6]	2.5	48	10[1,7]	2.5	48	0[1,7]
	2.0	44	0	2.5	60	13	2.5	51	0
							2.5	60	0
							2.5	48	0
Wooden piece b[4] (50 × 50 × 100 mm, *hinoki*, density 0.46 g/ml)	2.0	53	0[6]	2.5	58	7[1,8]	2.5	58	0[1,8]
	2.0	50	0	2.5	55	11	2.5	61	0
							2.5	55	0
							2.5	82	0
Wooden piece c[3] (65 × 65 × 200 mm, *hinoki,* density 0.65 g/ml)	3.8	99	0[6]	5.0	61	11[1,7]	2.5	61	0[1,7]
	3.8	118	0	5.0	55	13	2.5	55	0
Wooden piece d[5] (100 × 100 × 200 mm, Japanese cedar, density 0.38 g/ml)	10	447	0[6]	10	447	0[6]	–		
	10	375	0	10	375	0			
Wooden piece e[5] (100 × 100 × 200 mm, Japanese cedar, density 0.38 g/ml)	10	437	0[7]	10	437	0[7]	–		
	10	407	0	10	407	0			
Silicon lid with a capillary tube (1 mm diameter)	3.8	26	0	–			–		
	3.8	43	0	–			–		

[1] Emergence of adult insects (viable eggs, larva and pupa) 8 weeks after treatment

[2] Rice, with eggs, larva and pupa inside, was put into a hole (20 × 20 × 20 mm) bored in the centre of the books. The books were put tightly inside a paper box

[3] Rice, with eggs, larva and pupa inside, was put into a hole (12 mm diameter) bored halfway into the transversal section of a wooden piece

[4] The same as 3 except the hole was bored into the tangential section

[5] The same as 3 except the size of the hole is 20 mm diameter

[6] Uncoated wood

[7] Both sides of transversal section were coated with a silicon adhesive

[8] Both sides of tangential section were coated with a silicon adhesive

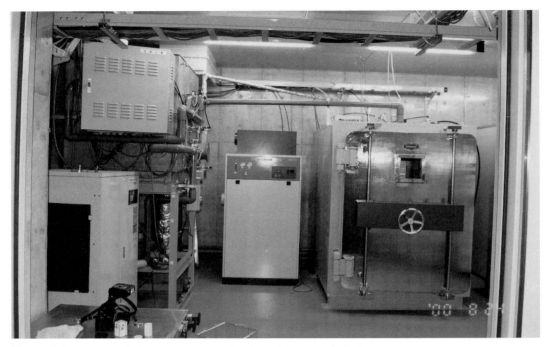

Figure 1 An automated nitrogen treatment chamber with an attached nitrogen generator

Research Institute for Cultural Properties (Kigawa *et al.*, 1999b) (Figure 1). It is an experimental system and the first of this type in Japan. The air-tight steel chamber is 3.0 m³ and equipped with temperature, relative humidity and oxygen control devices. In this system, the oxygen concentration fell below 0.1% within eight hours after the start of treatment. Although permanently fitted with a nitrogen-generating unit, the chamber can also be used with argon or carbon dioxide or any combination of these gases, by attaching appropriate gas supplies.

To eliminate *Lasioderma serricorne* from infested materials, the chamber was first used for plant specimens and then for the contents still in their drawers, but removed from the cabinet. The chamber system maintained the oxygen concentration near 0.08% at 30 ± 0.5°C and 55 ± 3% RH for a three-week period; the treatment condition for the most tolerant group A. In another case, feather, wool and skin objects, which were infested by varied carpet beetle, *Anthrenus verbasci* (group B), were treated with an oxygen concentration near 0.08% at 30 ± 0.5°C and 55 ± 3% RH for one week. Various developmental stages of black carpet beetle *Attagenus japonicus* were also treated at the same time. A 100% kill of the insects was confirmed in all treatments. The chamber has been easy to operate, even for a minimally-trained conservator, making this type of anoxic treatment system a safe and effective alternative for insect eradication in infested cultural objects.

RP system® (RP Agent-K type) oxygen scavenger

RP Agent-K type is an oxygen scavenger that does not absorb or discharge moisture (Mitsubishi Gas Chemical Company Inc., 1999). Because Ageless® Z discharges moisture, which is sometimes troublesome, this

characteristic of K type is quite useful. K type maintains existing moisture levels inside the bag, and so it can be used for organic materials such as wood, paper and textiles. In joint research with Ms Mutsumi Aoki of the National Institute of Japanese Literature and the Mitsubishi Gas Chemical Company Inc., we performed treatments with RP Agent-K type to treat ancient manuscripts and hanging scrolls (Figure 2) which were infested by book-borer anobiid and pubescent anobiid.

Carbon dioxide treatments

For carbon dioxide treatments, we have used treatment tents made of a barrier film. Although we do not humidify the carbon dioxide gas, it seems there are no serious problems with treated materials. In the joint study with staff from the Centre for Conservation Science, Gangoji Institute for Research of Cultural Property, we performed carbon dioxide treatment of various kinds of folk museum objects and ancient manuscripts. Treatment was carried out in a tent made of a barrier film which has a very low permeability to carbon dioxide (4 ml/(m² × days × atm.)). A two-week treatment with carbon dioxide was successfully performed without additional purge of the gas during the whole treatment. Therefore, carbon dioxide may be the gas of choice to treat large batches of such kinds of objects with the cooperation of pest control companies.

CONCLUSIONS

From the results of our experiments, we have devised a protocol of low oxygen and carbon dioxide treatments to facilitate practical eradication of insect pests in cultural institutions in our country. Infested objects should be exposed to low oxygen levels at 25°C or 30°C for one to

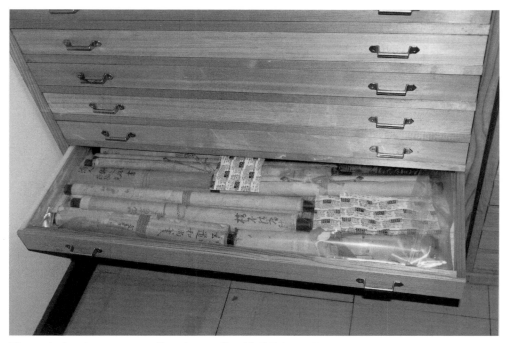

Figure 2 On-site treatment of hanging scrolls with RP Agent K-type oxygen scavenger

three weeks depending on the insect species. At 20°C, ten weeks exposure is needed. Exposure to carbon dioxide should be for one or two weeks at 25°C depending upon the species. For insects that tunnel deep inside books and wooden objects, the rate of gas exchange inside the pieces varied with the density, size, or coating materials of the object, making it difficult to determine a general treatment condition. An automated fumigation chamber with an attached nitrogen generator was shown to be effective and practical for treating small objects. Also effective were oxygen scavengers, such as RP system® (RP Agent-K type). Carbon dioxide treatments in barrier-film tents are probably more effective for large-scale treatments.

REFERENCES

Bailey S W, Banks H J, 'A review of recent studies of the effects of controlled atmospheres on stored product pest', in *Controlled Atmosphere Storage of Grains,* Shejbal J (Editor), 1980, Elsevier, Amsterdam, 101–118.

Bell C H, 'Effects of temperature on the toxicity of carbon dioxide to insects', in *Controlled Atmosphere and Fumigation in Grain Storages,* Ripp B E (Editor), 1984, Elsevier, Amsterdam, 67–75.

Bell C H, 'Alternatives – physical methods and emission reduction', in *Agrochemicals and Plant Protection,* Vol.1, *The Methyl Bromide Issue,* Bell C H *et al.* (Editors), 1996, John Wiley and Sons, 323–389.

Gilberg M, 'Inert atmosphere fumigation of museum objects', in *Studies in Conservation,* 1989, **34**, 80–84.

Gilberg M, 'Inert atmosphere disinfestation using Ageless oxygen scavenger', in *ICOM Committee for Conservation 9th Triennial Meeting,* 1990, Dresden, 812–816.

Gilberg M, 'The effects of low oxygen atmospheres on museum pests', in *Studies in Conservation,* 1991, **36**, 93–98.

Kigawa R, Miyazawa Y, Koizumi M, Sano C, Miura S, Nochide H, Kimura H, Tomita B, 'Evaluation of the effects of various pest controlling reagents on pigments and metals: effects of pesticides, fumigants, carbon dioxide and nitrogen', in *Kobunkazai no kagaku,* 1999a, **43**, 12–21.

Kigawa R, Yamano K, Miura S, Maekawa S, 'An automated anoxic treatment chamber for cultural objects, a novel system with nitrogen generator', in *Science for Conservation,* 1999b, **38**, 1–8.

Kigawa R, Yamano K, Miura S, Jippo K, Miyazawa Y, Maekawa S, Nochide H, Kimura H, Tomita B, 'Low oxygen atmosphere and carbon dioxide treatments for eradication of insect pests in Japan', in *Pre-print of Integrated Pest Management in Asia for Meeting the Montreal Protocol, the 23rd International Symposium on the Conservation and Restoration of Cultural Property,* 1999c, Tokyo National Research Institute of Cultural Properties, Tokyo, Japan, 63–72.

Kigawa R, Nagayama A, Yamano K, 'Effect of temperature on *Sitophilus zeamais* mortality in low oxygen treatment', in *Science for Conservation,* 1999d, **38**, 9–14.

Mitsubishi Gas Chemical Company Inc., *Manual of Cultural Objects Protection from Insect Pests by RP-K Type,* Japan, 1999.

Newton J, Abey-Koch M, Pinniger D B, 'Controlled atmosphere treatment of textile pests in antique curtains using nitrogen anoxia – a case study', in *Proceedings of the 2nd International Conference on Insect Pests in the Urban Environment,* Wildey K B (Editor), 1996, Heriot-Watt University, Edinburgh.

Reichmuth C, 'Low oxygen content to control stored product insects', in *Proceedings of the 4th Working Conference on Stored-Product Protection, 1986,* Donahaye E and Navarro S (Editors), 1987, Bet Dragon, Agricultural Research Organization, Tel-Aviv, Israel, 194–207

Rust M K, Kennedy J M, *The Feasibility of Using Modified Atmospheres to Control Insect Pests in Museums,* 1993, GCI Scientific Program Report, The Getty Conservation Institute, Los Angeles.

Selwitz C, Maekawa S, *Inert Gases in the Control of Museum Insect Pests,* 1998, The Getty Conservation Institute, Los Angeles.

Smith C P, Newton J, 'Carbon dioxide: The fumigant of the future', in *International Seminar on Research in Preservation and Conservation,* Saur K G (Editor), 1991, International Federation on Library Associations, Columbia University, New York.

Soma Y, Kishino H, Goto M, Yabuta S, Matsuoka I, Kato T, 'Response of stored grain insects to carbon dioxide 1. Effects of temperature, exposure period and oxygen on the toxicity of carbon dioxide to *Sitophilus zeamais* Motschulsky, *Sitophilus granarius* L. and *Tribolium confusum* Jaquelin du Val', in *Res. Bull. Pl. Prot.* 1995, Japan, **31**, 25–30.

Umney N, 'Low oxygen atmosphere for insect pest control in furniture, selection, investigation, application', in *Pest Attack and Pest Control in Organic Materials,* Neher A and Rogers D (Editors), 1997, Postprint of the UKIC furniture section conference, Museum of London, 1996, UKIC, 20–33.

Valentin N, 'Comparative analysis of insect control by nitrogen, argon and carbon dioxide in museum, archive and herbarium collections', in *International Biodeterioration and Biodegradation,* 1993, **32**, 263–278.

MATERIALS AND EQUIPMENT

RP system® (RP Agent-K type) and Ageless® Z oxygen scavengers

> Mitsubishi Gas Chemical Company Inc
> Information and Advanced Materials Company
> Oxygen Absorbers Division
> 2-5-2, Marunouchi
> Chiyoda-ku
> Tokyo 100-8324
> Japan
> Fax: +81 332 83 5187

Nitro-system (Nitrogen generator with automated humidification system)

> Kojima Instruments Inc.
> 14-1, Tachikawa-shiogatani
> Ujidawara-cho
> Tsuzuki-gun
> Kyoto 610-0231
> Japan
> Fax: +81 774 88 5525

ACKNOWLEDGEMENTS

We appreciate Mr Shin Maekawa of the Getty Conservation Institute for helpful advice in our planning of the automated nitrogen treatment chamber. We also appreciate the co-operation of Ms Mutsumi Aoki of the National Institute of Japanese Literature, Mr Shingo Hidaka, Ms Hitomi Date and Mr Fumitake Masuzawa of the Centre for Conservation Science, Gangoji Institute for Research of Cultural Property, Mr Masahiko Tanaka, Mr Tomoharu Himeshima and Kazuyuki Tomita of the Mitsubishi Gas Chemical Company Inc. in the joint studies. We also thank Ms Kazuko Jippo, Akiko Isshi and Mizuho Higuchi for their technical support. This work is supported by a grant from the Agency for Cultural Affairs, Japan and a grant-in-aid for scientific research from the Ministry of Education, Science and Culture, Japan.

BIOGRAPHY

Rika Kigawa received a PhD in the field of molecular biology, yeast genetics, from the Department of Biophysics and Biochemistry, Faculty of Science, University of Tokyo, in 1993. She started work as a researcher of the Biology Section, Conservation Science Department, Tokyo National Research Institute for Cultural Properties in 1993. Since 1999 she has been senior researcher. Her main research is on control of biodeterioration by insects, fungi and bacteria.

CHAPTER FOURTEEN

Nitrogen treatment: an insect case study

Monika Åkerlund

Swedish Museum of Natural History, Box 50007, SE-104 05 Stockholm, Sweden
Tel: +46 8 51954201   Fax: +46 8 51954085   e-mail: monika.akerlund@nrm.seand

AND

Jan-Erik Bergh

Dalarna University, Campus Falun, SE-791 02 Falun, Sweden
Tel: +46 23 778373   Fax: +46 23 778082   e-mail: jeb@du.se

Abstract

Anoxic treatment was tested for pest control on three types of insect cases. The test specimens included larvae of *Trogoderma angustum* (Solier) and *Anthrenus verbasci* (L). The cases were enclosed in low diffusion plastic and the oxygen level was reduced by use of the VELOXY® nitrogen generator. The following tests were performed. (1) High levels of nitrogen gas for one week. (2) High levels of nitrogen gas for two weeks. (3) Nitrogen gas with 2 sachets of Ageless® Z–200 added, with one-week exposure time. (4) Nitrogen gas with 2 sachets of Ageless® Z–200 added, with two-week exposure time. (5) Nitrogen gas with 4 sachets of Ageless® Z–200 added, with one-week exposure time. (6) Nitrogen gas with 8 sachets of Ageless® Z–200 added, with one-week exposure time. The oxygen levels were reduced to 0.2% by the VELOXY®. The temperature in the cases remained at approximately 22°C with a relative humidity (RH) level of approximately 28%. In the treatments using VELOXY® only (1 and 2), all the test specimens died. Using 2 sachets of Ageless® Z–200 (treatments 3 and 4), almost all the insects survived after one-week exposure, whereas after two weeks of exposure only a few larvae survived in the most airtight case. In treatment (5) using 4 sachets of Ageless® Z–200, 3 *Trogoderma* survived in the most airtight case. Using 8 sachets of Ageless® Z–200 (treatment 6) all the insects died.

Keywords

Anoxic, nitrogen generator, oxygen scavenger, insect case, entomological collection, pest control, *Trogoderma angustum*, *Anthrenus verbasci*

Introduction

Entomological collections are frequently attacked by pest insects. Dead insects are the natural food for pest species like dermestide beetles (*Anthrenus, Attagenus, Reesa, Trogoderma*), and moths (*Tineola, Tinea*). As the items are small, a pest infestation will have a devastating effect on the collection in a very short time. Rigorous pest control is therefore essential. The pesticides previously used for protection (paradichlorobenzene, DDT and dichlorvos) are now banned in many countries. Freezing, the most common alternative method, also has disadvantages.

The pinned insects are usually kept in special wooden storage cases with glass lids. The freezing treatment can cause the glass to break, old pins to corrode and items glued to paper to come loose. Therefore, another method developed for museum use in the last two decades, anoxic treatment (replacement of oxygen by, for example, nitrogen) (Daniel *et al.*, 1993), would be of value for pest

control of entomological collections. The insect cases to be treated are constructed with tight seals to prevent pest insects from entering. This could be an obstacle for the effectiveness of an anoxic treatment. The aim of the present study was to find out whether the oxygen concentration inside the closed case could be decreased to a level low enough to kill pest insects, and, if so, whether it could be done in a reasonable time.

Material and methods

The VELOXY® system

VELOXY®, developed under the European project SAVE ART, is a nitrogen generator system for anoxic treatment of museum objects to control insect pests. It produces an almost oxygen free (< 0.1% O_2) flow of 200 L/h that is used to modify the atmospheric composition, inside enclosures tailor-made from the gas-proof plastic barrier

Table 1 Experimental details

Case number	Treatment	Exposure (weeks)
A1, B1, C1	VELOXY®	1
A2, B2, C2	VELOXY® + 2 Ageless® Z–200	1
A3, B3, C3	VELOXY®	2
A4, B4, C4	VELOXY® + 2 Ageless® Z–200	2
A5, B5, C5	Control	1 and 2
A6, B6, C6	VELOXY® + 4 Ageless® Z–200	1
A7, B7, C7	VELOXY® + 8 Ageless® Z–200	1
A8, B8, C8	Control	1

film. The nitrogen generator is based on the use of tiny semipermeable hollow fibres that are connected to a manifold to which pressurized air is applied. Oxygen permeates through the walls of the fibres leaving a stream of nitrogen and the minor constituents of air. The permeation system is 37 cm wide, 40 cm long and 94 cm high and has wheels to make it portable.

The VELOXY® method uses a compressor and in these experiments a 1.1 kW unit was used which operated on the normal domestic electricity supply. It was 70 cm long, 26 cm wide, and 83 cm high, and it also has wheels and is portable.

For the tests, three types of wooden cases with glass lids were used. All the cases were 6 cm × 40 cm × 43 cm. Type A was an old case with one groove, type B was a new case with one groove and type C was a new case with two grooves. The details of the experiments performed in this work are given in Table 1.

Pest species used in the tests

Larvae of two pest species were used in the tests: *Trogoderma angustum* and *Anthrenus verbasci*. *Trogoderma angustum* (Solier) has spread from South America to North America and Europe. The larval stage can damage objects of animal origin as well as plant origin. *Anthrenus verbasci* (L), the varied carpet beetle, is more or less a cosmopolitan. The larvae are severe pests of wool, leather, fur, hair, etc. (Mroczkowski, 1968; Philipp, 1968; Åkerlund, 1991; Gerozisis and Hadlington, 1995).

The insects used in our experiments had been reared at the Danish Pest Infestation Laboratory. The larvae were placed within the cases in vials covered with cotton gauze, with 20 larvae in each vial. A paper box with pinned insects was also placed in the case. The case was enclosed in low diffusion plastic (RGI®),[1] which was heat sealed by the use of a Hawo sealer.[2] The oxygen level was reduced by use of the VELOXY® nitrogen generator and the gas was humidified by a Rentokil® humidifier.

The oxygen level in the enclosures was measured by an oxygen analyser (Analox Oxygen Analyser)[3] connected to the outlet valve.

Experiment 1

The first experimental set of cases A1, A3, B1, B3, C1 and C3 (Table 1) were treated with nitrogen only, whilst the second set, A2, A4, B2, B4, C2 and C4 were treated with nitrogen and 2 sachets of Ageless® Z–200. First, the vials with the test insects were placed horizontally in one corner of the case. The cases were then enclosed in low diffusion plastic and the oxygen level was reduced to 0.2% by the VELOXY®. Plastic tubes, with an inside diameter of 7 mm, connected the device to the enclosure. The gas entered as a gentle flow through one valve and left the enclosure through a non-return valve.

The reduction of oxygen inside the cases was due only to the diffusion of the gas. At the start of the treatment the cases were connected to each other through tubes and the non-return valve was placed at the end of the series of cases. Once the oxygen level had decreased to about 5%, the cases were treated one at a time. In the cases where Ageless® Z–200 was used, the enclosures were first cut open and then the Ageless® Z–200 sachets were added. The hole in the enclosure was then immediately heat-sealed.

The cases which were treated with VELOXY® only were separated from the nitrogen flow during the night and at weekends. The oxygen level, which then increased to between 0.4 and 3.4%, was actively reduced to 0.2% during the day. Case B3 (Table 2) had a leakage during the first night and with this case only, the enclosure was resealed and the treatment was restarted on day 1 at midday. The temperature was approximately 22°C but had a peak of 24°C on day 2. The RH level in the room was 15–30% and the nitrogen gas was conditioned to about 50% RH. The experiments were carried out for one and two weeks, respectively. Controls, A5, B5 and C5, were set up in an identical way, but without the nitrogen treatment.

Experiment 2

In the second experiment the same method was used on cases A6, A7, B6, B7, C6 and C7, with the exception that 4 or 8 sachets of Ageless® Z–200 were added and a humidity logger (Mätman datalogger) was placed inside two of the cases. The vial with the test insects was placed in an upright position in one corner of the case. The exposure time was one week. The oxygen level was reduced to 0.2%. The room temperature was 22–24°C and the RH level in the cases was approximately 28%. Controls, A8, B8 and C8, were set up in an identical way, but without the nitrogen treatment.

Table 2 Oxygen level in the enclosures measured in the mornings

Case number	Day 1 O_2 (%)	Day 4 O_2 (%)	Day 5 O_2 (%)	Day 8 O_2 (%)	Day 13 O_2 (%)	Day 14 O_2 (%)
A1	0.6	0.4	0.2	0.2	–	–
A3	1.4	0.4	0.3	0.2	0.2	0.3
B1	1.4	0.6	0.3	0.4	–	–
B3	⋆	0.8	0.2	0.2	0.2	0.2
C1	2.7	0.7	0.3	0.2	–	–
C3	3.4	1.3	0.4	0.5	0.3	0.4

⋆ not measurable, enclosure resealed, treatment restarted and O_2= 0.2% at midday

After the treatment, the condition of the larvae was checked three times during one week. This entailed touching the larvae with a pin, gently exhaling towards them and observing their movements under a stereo microscope.

RESULTS

As the vials with the test insects were placed horizontally in the first test, some test insects had escaped and had to be recollected. A few of them were not found. The number of test insects (*n*) in Tables 3, 4, 5 and 6 are the recollected ones and those that remained in the vial.

For the type of cases used in Experiment 1, which were treated with an oxygen scavenger, an oxygen level of 0.2% in the enclosure at the start and 2 sachets of Ageless® Z–200 for one week, the treatment was not effective in killing the insect pests.

For the cases of type A and B, with one groove, treatment with 4 sachets of Ageless® Z–200, killed both insect species in one week, and took two weeks when 2 sachets were used.

For the cases of type C, with two grooves, one-week treatment with 4 sachets of Ageless® Z–200 was enough to kill the larvae of *Anthrenus verbasci*, but two weeks were needed to kill the *Trogoderma angustum*. In the treatment with 8 sachets of Ageless® Z–200, all larvae of both species were killed within a week.

The one-week exposure treatment with VELOXY® was enough to kill all the larvae of both species. After one week of treatment, the oxygen levels in the enclosures A1 and C1 were 0.2% and for B1 0.4%. After two weeks, the mortality was 100% and the enclosures had an oxygen level ranging from 0.2% to 0.4%.

Table 3 One-week treatment with VELOXY® and 2 sachets of Ageless® Z–200

Case number	O_2 at end of treatment (%)	Test species⋆	*n*	Survival rate (%)
A2	1.8	T. angustum	20	100
		A. verbasci	20	0
B2	0.6	T. angustum	20	95
		A. verbasci	20	60
C2	19.3	T. angustum	20	85
		A. verbasci	19	53

⋆ T. angustum = *Trogoderma angustum*; A. verbasci = *Anthrenus verbasci*

Table 4 Two weeks treatment with VELOXY® and 2 sachets of Ageless® Z–200

Case number	O_2 at end of treatment (%)	Test species⋆	*n*	Survival rate (%)
A4	0.4	T. angustum	20	0
		A. verbasci	20	0
B4	0.4	T. angustum	20	0
		A. verbasci	20	0
C4	1.0	T. angustum	20	5
		A. verbasci	20	5

⋆ T. angustum = *Trogoderma angustum*; A. verbasci = *Anthrenus verbasci*

Table 5 Control after one and two weeks for Experiment 1

Case number	Test species★	n	Survival rate (%)	
			One week	Two weeks
A5	*T.angustum*	16	94	94
	A.verbasci	17	94	94
B5	*T.angustum*	20	100	80
	A.verbasci	19	95	79
C5	*T.angustum*	19	95	79
	A.verbasci	20	100	50

★ T. angustum = *Trogoderma angustum*; A. verbasci = *Anthrenus verbasci*

Table 6 One-week treatment with VELOXY® and 4 sachets of Ageless® Z–200

Case number	O_2 at end of treatment (%)	Test species★	n	Survival rate (%)
A6	0.2	*T.angustum*	17	0
		A.verbasci	18	0
B6	0.2	*T.angustum*	19	0
		A.verbasci	17	0
C6	0.2	*T.angustum*	14	21
		A.verbasci	17	0

★ T. angustum = *Trogoderma angustum*; A. verbasci = *Anthrenus verbasci*

DISCUSSION

In the one-week treatment with VELOXY® and 2 sachets of Ageless® Z–200, the survival rate of the test insects was high but the result was rather confusing. In the old case, all larvae of *Trogoderma angustum* survived, but all the larvae of *Anthrenus verbasci* died. We have no obvious explanation for that result. The survival of the test insects in case C2, the most airtight of the cases tested, can be explained by the fact that the oxygen level was 19.3%. The enclosure obviously had a leak. The survival of the insects, however, was less than in case B. We also have no explanation for that result.

The survival in the control experiment was high after one week, but rather lower after two weeks. The low humidity of the air may be a stress factor for the larvae. The test insects in the control had no access to food during the second week and they had obviously consumed some of their companions.

The oxygen analyser used in the tests demands a flow of gas. If the amount of gas in the enclosure is not sufficient, the device might not be able to detect the correct gas concentration. The analyser might not be completely accurate at low levels of 0.1–0.2.

The results of the tests can only be referred to the particular conditions that existed during the tests. The number of replicates in the tests was very small and more tests are needed to get a satisfactory result.

The treatment with VELOXY® only was efficient, but demanded very intensive work, by checking the oxygen level in each enclosure and keeping the level down. It is possible to have the device keeping a constant flow of nitrogen during the entire treatment. However, that would occupy the equipment and prevent treatment of other objects during that time. Using both the nitrogen generator and oxygen scavenger is a more efficient method for museum staff.

Due to the fact that the vials were placed horizontally, some of the first test insects escaped and had to be recollected. The fugitives were found inside the pinned insects, in the unit box in the case, and for cases of type A and B, at the groove between the case and the lid, and outside the case in the enclosure. Only cases of type C were tight enough to prevent the larvae from leaving. This implies that the type C case would also prevent insects from entering.

CONCLUSIONS

It is possible in practice to use the VELOXY® method for pest control of closed wooden insect cases with glass lids. Adding enough of the oxygen scavenger to the enclosures instead of only using the VELOXY® reduces the work input. Care must be taken to determine the construction of the case, as a tighter case needs a longer exposure time.

NOTES

[1] The plastic film was nylon and polyethylene 0.09 mm thick. Oxygen permeability 0.45 cm^3 mil/(m^2 × day × atm.).

Table 7 One-week treatment with VELOXY® and 8 sachets of Ageless® Z–200

Case number	O₂ at end of treatment (%)	Test species*	n	Survival rate (%)
A7	0.2	*T.angustum*	19	0
		A.verbasci	20	0
B7	0.2	*T.angustum*	18	0
		A.verbasci	20	0
C7	0.1	*T.angustum*	18	0
		A.verbasci	19	0

★ T. angustum = *Trogoderma angustum*; A. verbasci = *Anthrenus verbasci*

Table 8 Control after one week for Experiment 2

Case number	Test species*	n	Survival rate (%)
A8	*T.angustum*	17	100
	A.verbasci	16	100
B8	*T.angustum*	15	100
	A.verbasci	10	100
C8	*T.angustum*	18	100
	A.verbasci	17	100

★ T. angustum = *Trogoderma angustum*; A. verbasci = *Anthrenus verbasci*

[2] The Hawo heat sealer type hpl WSZ-300. 360W. Seal length 0.3 m, width 11 mm.

[3] Analox 101D2 and Analox oxygen sensor Type 9212-5A Scottish Anglo Environmental Protection Ltd Accuracy ±1%.

REFERENCES

Åkerlund M, 'Ängrar – finns dom...? Om skadeinsekter i museer och magasin', 1991, Uppsala, Svenska museiföreningen.

Gerozisis J, Hadlington P, *Urban Pest Control in Australia*, 3rd edition, 1995, New South Wales University Press, Sydney.

Daniel V, Hanlon G and Maekawa S, 'Eradication of insect pests in museum using nitrogen', in *WAAC Newsletter*, 1993, **15**(3), 15–19.

Mroczkowski M, 'Distribution of the *dermestidae* (Coleoptera) of the world with a catalogue of all known species', in *Annales Zoologici*, 1968, **3**, 1–177.

Philipp E, 'Zur kenntnis der morphologie und biologi von *Trogoderma angustum* Solier, 1849 (Coleoptera Dermestidae)', in *Zeitschrift für Angewandte Zoologie*, 1968, **55**(2), 193–256.

MATERIALS AND EQUIPMENT

Humidity logger
Eltex of Sweden AB
Box 608
SE-343 24 Älmhult
Sweden
Tel: +46 476 488 00

Eltex (UK) Limited
Lane Close Mills Bartle Lane
Great Horton
Bradford BD7 4QQ
United Kingdom
Tel: +44 1274 571071

Oxygen analyser and gas humidifier
Rentokil Initial Limited
Rentokil House
Garland Road
East Grinstead
West Sussex RH19 1DY
United Kingdom
Tel: +44 800 136668

Heat sealer
HAWO GmbH
Obere Aue 2–4
D-74847 Obrigheim
Germany
Tel: +49 6261 97700

VELOXY® system (incl. Nitrogen generator, gas humidifier, heat sealer, oxygen analyser, plastic film, valves plastic tubes)
RGI Resource Group Integrator s.r.l.
Via Nazario Sauro
816145 Genova
Italy
Tel: +39 010 3626002

Ageless® oxygen scavenger
   Mitsubishi Gas Chemical Europe GmbH
   Immermannstrasse 45
   (Deutsch – Japanische Centre)
   4000 Dusseldorf 2
   Germany
   Tel: +49 211 363080

Acknowledgements

Financial support has been given by the Commission of the European Communities, DG XII (Contract ENV4-CT98-0711), Swedish National Heritage Board (the project Oxygen Absorber in Practical Conservation), The Research Council of Arts and Education at Dalarna University, Sweden and the Swedish Museum of Natural History. We thank our colleagues at the Danish Pest Infestation Laboratory for helpful assistance and the members of PRE-MAL for valuable and constructive criticism.

Biographies

Monika Åkerlund BSc. is an entomologist at the Swedish Museum of Natural History and the secretary of the Swedish working group PRE-MAL (Pest Research and Education – Museums Archives and Libraries). She is working on research, education and consulting in pest control in museums.

Dr Jan-Erik Bergh is an entomologist and senior lecturer at Dalarna University. He is the project leader, responsible for the research of PRE-MAL. His main research interest is low oxygen treatment.

CHAPTER FIFTEEN

CARBON DIOXIDE FUMIGATION: PRACTICAL CASE STUDY OF A LONG–RUNNING SUCCESSFUL PEST MANAGEMENT PROGRAMME

Sue Warren

CONSERVATOR

Canada Science and Technology Museum, 1867 St. Laurent Blvd, PO Box 9724, Ottawa Terminal, Ottawa, Ontario K1G 5A3, Canada

Tel: +1 613 991 3061   Fax: +1 613 991 0827   e-mail: swarren@nmstc.ca

ABSTRACT

The Canada Science and Technology Museum has been using carbon dioxide fumigation in their pest management programme since 1991. We were one of the first museums in North America to use this type of fumigation system and the first in Canada. Through ten years of practical experience we have proven that carbon dioxide fumigation is not only effective, but also economical and user-friendly. We have successfully eliminated what was a very active clothes moth infestation and have implemented an ongoing plan to monitor and control pests in our diverse collection. Based on our success with controlled CO_2 tests on infested vehicles and with colonies of live moths and larvae, we purchased our system in 1992 with an initial investment of $12,000 Canadian. In addition to active fumigation, the current pest management regime includes an assessment programme to identify objects which are, or could possibly be, infested by creating a log of artefacts in need of fumigation. There are annual evaluations for re-infestation or recurrence of infestation, with special attention given to examining those vehicles or objects that were particularly active. This paper will discuss the scope of the problem, with emphasis on dealing with large objects. This is followed by a description of the preventive and interventive measures which have been implemented in order to eradicate the pest problem, and to prevent future infestations. A description of the exact methodology will be given, and recommendations based on our experiences to help avoid mistakes and failures.

KEYWORDS

Carbon dioxide, fumigation, controlled atmosphere, case-bearing clothes moth, pest management

INTRODUCTION

Theory of carbon dioxide fumigation

It is generally accepted that the use of inert gases to control insect pests works through the mechanisms of oxygen deprivation and desiccation. Insect physiology is adapted to control the exchange of oxygen and carbon dioxide, and to conserve moisture through the same respiratory system. Decreasing the available oxygen, and replacing it with carbon dioxide, essentially strains the insect's gas exchange system, requiring it to process more 'air' in order to gain sufficient oxygen; thereby losing valuable moisture.

Critical levels of CO_2 for fumigation have been determined through experience in the food industry and recently in the conservation profession. It has been proven that a 60% CO_2 level will achieve 100% mortality of all stages of insect life; and that mortality rates increase as the temperature rises and humidity falls (Selwitz and Maekawa, 1998). The range of times for total mortality lies somewhere between 10 and 21 days, depending on temperature conditions and species of pest.

BACKGROUND

The Canada Science and Technology Museum has been using carbon dioxide fumigation for the treatment of artefacts since June 1991. After ten years we have ample evidence that the system is effective for the eradication of insect pests, that it is an economical means of control and that, with adequate precautions, it is easy and safe to use for both the artefacts and the operators.

We were the first Canadian institution to put CO_2 fumigation into practice, although not the first museum to use this system. Several institutions in the United States were experimenting with CO_2 fumigation in the years immediately preceding our purchase. Our system was continually in use throughout the warm weather months during the first six years of operation. It is now used periodically as a preventive measure when new acquisitions are thought to pose a threat or when a problem is identified in a warehouse through regular inspections.

IDENTIFYING THE PROBLEM

The conservation division was established in 1985, and represented quite a change from previous restoration and preservation strategies at the museum. One of the first priorities of the new division was to assess and improve the storage facilities. There were four warehouses in use at the time, and a large area of outdoor storage. Figure 1 shows the agricultural artefact warehouse, which presents a huge potential for infestation. Two of the warehouses in particular presented problems with pest control. First was the agricultural artefact warehouse, which was playing host to a range of pests from insects to rodents. Second was the so-called land transportation warehouse containing the rail cars, street cars, some automobiles and the horse-drawn vehicle collection, all of which were infested with case-bearing clothes moths, *Tinea pellionella*. Since the latter warehouse was scheduled for renovation, it became the starting point for the pest control programme.

The carriages were initially blamed for the infestation; having been transferred from the collections of several other institutions and being in very poor condition. Figure 2 illustrates characteristic damage to trimmings of the horse-drawn vehicles. It became apparent from the degree

of damage and the high activity level that the infestation had been active in the carriages for many generations of insects. With subsequent inspections, *Tinea pellionella* was found to be equally active in some of the street cars and rail cars.

Some of the museum staff, who had been at the institution for many years, remembered having the warehouse fumigated 'years ago' so it was obvious that a problem had been identified at some point. There were 'Vapona' dichlorvos slow-release fly strips hanging in some of the automobiles and rail cars; but these had long-since turned to sticky brown messes, which dripped onto carpets and seats. Nobody remembered who had installed the strips, or what fumigant had been used for the warehouse treatment. Apparently there was no follow-up. A rather desultory assessment of the carriages, done in the early 1980s, noted that the carriages were dirty and needed to be vacuumed.

We began looking into methods of fumigation in 1988. It was obvious that we were well past the 'good housekeeping' stage. Chemical fumigation was not even considered, due to the cost of installing facilities and the need for licensing. We required a system which could be done in-house, and which could accommodate large vehicles and agricultural equipment, as well as smaller objects. The large size of our artefacts and the number of infested objects made it uneconomical to rent local freezer facilities. Not only would the space requirement have been huge but also the cost of transporting the objects would have been horrendous. Despite preconceived notions about winter temperatures in Ottawa, the very cold season is not long enough, nor

Figure 1 Agricultural artefact warehouse

Figure 2 Characteristic damage to trimmings of the horse-drawn vehicles

reliably cold enough, to have made outdoor freezing a feasible option either.

In 1990, in consultation with Tom Strang of the Canadian Conservation Institute, we decided to experiment with carbon dioxide fumigation. While new to the museum community, CO_2 had been widely used in the food and agriculture industries, and was being experimented with at several museums in the United States. Tom had been working with the process on a small scale, and was quick to encourage us to be a test site. We discarded the idea of nitrogen anoxia at this time, mostly due to the size of the artefacts we were dealing with, and the cost and difficulty involved in maintaining such a highly-controlled environment on a large scale.

EARLY EXPERIENCE

Our first experiment involved making our own enclosure from polyethylene sheet. We knew that it would not seal in the gas completely, but hoped to be able to maintain 60% concentration through a constant flow. In the end, we were never able to reach 60% CO_2 in the bag, let alone maintain it. We quickly abandoned this idea and proceeded to the next step.

In June 1991, we rented a standard B & G mini-bubble (Rentokil, UK) that measured roughly 2.5 m^3. We monitored the gas concentration with a hand-operated

Draeger gas detector, which had a detection limit of 60%. This was by no means ideal, since we were attempting to maintain concentrations above 60%; however, we decided that this was acceptable for our trials. In truth, we probably overcompensated with very high concentrations in these first attempts.

The experiments were carried out on sleighs, which were easy to transport and fitted nicely into the small bubble. Each cycle contained a control beaker of ten live moths and ten live larvae, *Tinnea pellionella* (in a cloth matrix). At this time, it was relatively easy to gather live specimens from a variety of vehicles, one of which we dubbed our 'breeding colony' sleigh. Figure 3 shows our breeding colony sleigh, which was the source of most of our control specimens and was by far the most actively infested vehicle. The 'control' insects were placed high inside the bubble, and sealed in with a Gore-Tex cover over the beaker.

A total of eight sleighs were fumigated in the rented bubble and then monitored for insect activity over the course of several months (including the autumn peak activity period). No recurrence of activity was noted; so in 1992, we purchased a custom-made bubble from B & G, with dimensions of 2.5 m high, 2.75 m wide and 6 m long. Because of the large size, an additional filling port was installed. The system included a vacuum pump and hoses, to which we added a CO_2 monitor with remote sensor, a data logger with real-time connection, a fan to circulate the gas, and an ultrasonic humidifier with built-in humidistat. We also made a few minor adaptations. We installed metal pulleys on the plastic loops at the top of the bubble, so that it could be easily raised when suspended under a shelving unit. We also built ramps to fit over the seal, so that it would not be subjected to crushing when large vehicles were rolled over it. Finally, a heavy tarp was placed on the floor of the bubble to protect it from wear, and from the occasional motor oil leak. All of the wiring for electrical appliances and computer cables was run through an ABS plumbing cab that was drilled to accommodate the wires, and sealed with caulk.

In an attempt to reduce the amount of gas required, we experimented using space-filling weather balloons, which we filled with CO_2 gas and then inserted inside and underneath the vehicles to reduce the air volume. Unfortunately, the balloons did not hold the CO_2 for the duration of the cycle, and while this had no ill-effect as far as reducing the concentration inside the bubble, they could not be re-used in subsequent fumigation cycles and the idea was therefore abandoned.

To date, 103 horse-drawn vehicles have been fumigated, 6 automobiles, a piano, numerous dismantled seats, carpets, and cushions, and a variety of agricultural objects. The fumigation programme is now mostly preventive: new acquisitions are isolated and observed, and fumigated if there is any evidence of infestation. Previously infested objects are monitored and stored in bags. Inspections are

Figure 3 Infested sleigh: the source of most of the control specimens

carried out on an annual basis, usually in the spring when activity is likely to be highest.

METHODOLOGY

The only limiting factor to CO_2 fumigation in our institution has been temperature. In the past, the bubble has been located in warehouses which were not well heated during the winter months. This is a necessity during Canadian winters, since heating an uncontrolled warehouse can result in relative humidity levels of 15% or lower. Consequently, it was impossible to carry out fumigation treatments during the winter months. A minimum temperature of 20°C is required with better results being obtained at higher temperatures. The average temperature of our fumigation cycles is 22°C (Newton, 1990).

There is no internal framework to the B & G fumigation bubble. It has loops on the cover, from which it can be suspended. It has always suited our purposes to

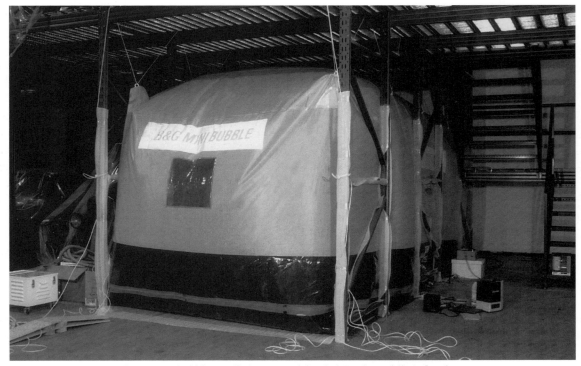

Figure 4 B & G mini fumigation bubble installed in one of the shelving bays, fully inflated

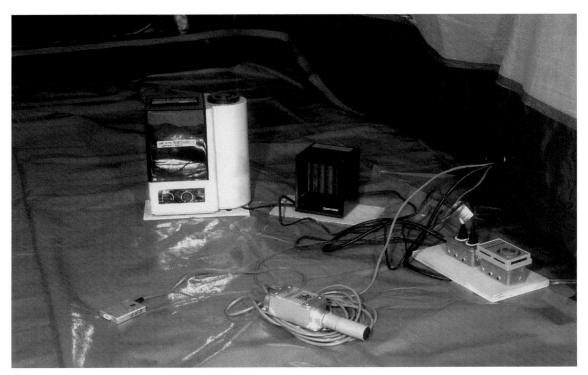

Figure 5 Gas sensor, humidifier, fan and data logger installed in the bubble

install the bubble in a shelving bay, and suspend the cover from the shelving frame above. Figure 4 shows the B & G mini bubble installed in one of our shelving bays, and fully inflated – in this case with air. This allows for easy access, particularly with large artefacts such as carriages, which can be rolled into the bubble while the cover is raised, and then have the cover simply lowered over them. Small objects are sometimes protected by a frame built around them so that they are not crushed by the weight of the deflated bubble. In some instances, a group of small objects can be covered by a table or box. All projections are padded with ethafoam, to prevent possible punctures to the membrane of the bubble. Doors and windows on the vehicles are opened, and fabrics are arranged as loosely as possible. In addition to the artefacts, an ultrasonic humidifier, data logger, and fan are also installed. Figure 5 shows the equipment installed in the bubble – gas sensor, humidifier, fan, and data logger.

The seam is sealed and the bubble is checked for air-tightness by inflating it with air overnight. The cycle begins in the morning, by evacuating all air from the bubble using the vacuum pump. The loose weight of the bubble cover is distributed over the artefacts, so that it does not snag or weigh too heavily in any one place. Once a vacuum has been drawn on the enclosure, two tanks of CO_2 are connected to separate ports, and the flow regulated to 70 standard cubic feet per hour (scfh). This rate was determined using two considerations: firstly, the filling of the bubble was planned to take place over the course of a working day, and secondly it would require no warming of the gas prior to entry into the chamber. Initially, we followed the rationale that it was better to introduce the

gas slowly so that the insects did not enter a state of stasis, whereby they may have been able to outwait the low oxygen levels. Whether or not this was valid, it seemed to work, and the slow fill time allowed the CO_2 to warm up merely by adapting to the large volume of warm air in the warehouse. I have never found it necessary to purge the bubble, as it was always possible to reach concentrations above 60% with one evacuation and filling. We still maintain this method of introducing CO_2.

The concentration of CO_2 usually attained at the end of the first day is somewhere in the region of 70–80%. This varies with the type of object inside: obviously a large enclosed coach will not be as completely evacuated as smaller open vehicles, so the maximum concentration of gas is lower at the start of the cycle. Maintenance throughout the cycle consists of a daily check to ensure that the concentration does not drop below 60%. Generally speaking, the concentration decreases by 2–3% on a daily basis, and will require 'topping up' intermittently. The humidifier will not require any maintenance, since it operates only during times when CO_2 is being introduced into the bubble. There has been disagreement about the use of humidification in CO_2 fumigation; obviously if desiccation is the aim, then humidifying the space will reduce the efficacy of the treatment. To counterbalance this, I have used humidification when treating horse-drawn vehicles, because of the materials and construction of the artefacts. Large panels of thin wood, expanses of leather under tension, and fragile painted finishes over wood, are all vulnerable to changes in humidity. The large volume of air and gas surrounding these materials means that there is no buffering effect (as occurs when a group of small

Figure 6 Part of the horse-drawn vehicle collection, fumigated and bagged in the storage area

densely packed artefacts is fumigated in a small air volume). The bubble is not usually humidified when there is a high density of artefacts relative to the air volume.

The duration of treatment depends on the type of object and the variety of pest. Generally, for clothes moths, the object remains in the bubble for 14 days and the treatment time for dermestids and woodboring insects is 21 days. Both of these treatment durations have been shown to be effective for our collection.

At the end of the required time period, the vacuum pump is used to evacuate the CO_2 from within the bubble, and vent it outside. The bubble is then inflated with air. This process may need to be done a second time in order to reduce the CO_2 level to a safe limit of less than 5%. The objects are then thoroughly vacuumed and where possible an indicator, such as a sheet of paper, is strategically placed underneath the object to catch frass and show any new insect activity and the artefacts are sealed in polyethylene bags. Figure 6 shows part of the horse-drawn vehicle collection, fumigated and bagged in the storage area. These bags are not intended to be airtight (in fact they never are) but they do control the spread of insect pests. The bags actually serve as a good indicator of activity, since moths and debris tend to accumulate in the loose corners and folds of the polyethylene.

PEST MANAGEMENT PROGRAMME

A comprehensive plan was initiated in 1993, to assess the condition of all the artefacts in the collection. This entailed completing a brief checklist for all artefacts: much shorter and easier to complete than a condition report and requiring no photo documentation. One of the categories

included in the form is 'Infested' with possible entries of 'yes', 'no' or 'unknown'. The category is searchable on our database, so that a list of problem artefacts can be generated. This has helped to identify pieces in need of fumigation, or at least in need of monitoring. Having been identified, the pieces are isolated (or bagged) to await a batch fumigation. This has been carried out, to date, by locating the fumigation bubble in different warehouses, and treating the objects from that particular warehouse before moving to the next site. All of our three current warehouses have been assessed and the objects fumigated, with the exception of large motorized vehicles, of which seven have been identified as having evidence of insect activity and five require monitoring.

One item that recently slipped by our screening process was an IBM typesetter keyboard, dating from the 1970s. It did not seem a likely candidate for infestation as it was fabricated of metals and polymers. What we found, however, was a fascinating example of webbing clothes moth (*Tineola bisselliella*) adaptation; the moths were living in the foam padding inside the machine. Foam does not usually satisfy the nutritional requirements of the webbing clothes moth and there were no natural fibres for them to consume. The moths were probably living off dirt gathered on the foam padding. However, it was interesting to note that accumulations of frass consisted of sticky black 'pellets' – obviously the digested by-product of the foam. Figure 7 shows webbing clothes moths inside the typesetter.

The typesetter was received in 1996 and was thought to smell bad, but otherwise ignored and stored in the new acquisitions storage area awaiting cataloguing. It languished for almost three years before being examined for

Figure 7 Webbing clothes moth inside an IBM typesetter

cataloguing. During this time, it could potentially have infested over 150 other artefacts that were stored nearby. Once the problem was identified, a list of all the artefacts known to have been in close proximity was generated and each piece was checked. This had the potential for a widespread infestation; but fortunately it proved to be less of a problem than imagined. No evidence was found in the storage area, and none of the other artefacts had been infested. Apparently the moth colony had died out prior to the acquisition. This was an important lesson and has encouraged us not to make assumptions about likely and unlikely candidates for infestation.

Yearly inspections of previously fumigated objects are undertaken, to monitor any recurrence of activity. More thorough inspections are carried out on objects that were most actively infested. Keeping the storage area clean, and having indicator papers under objects, helps with inspections. The polyethylene bags are problematic because they get torn and they make it difficult to see the objects. However, they are a necessary precaution not only to contain a recurrence of infestation, but in the case of the carriages to prevent re-infestation.

CONCLUSIONS

Every museum collection has the potential for infestation and it is important that this is accepted and precautions taken to prevent infestation occurring. When it does occur, it must be acknowledged and dealt with quickly and efficiently. Any commercial toxic fumigant gas poses a threat to workers and the public, and therefore should be avoided where alternatives exist. Facilities constructed for fumigation or freezing are prohibitively expensive and often not within the realm of possibility for small institutions. Controlled atmosphere bubbles, specifically those designed for the use of CO_2 gas, are relatively inexpensive to buy, are portable, can be operated by non-certified users (at least in Canada and the United States), and with adequate precautions, do not pose an environmental threat should a malfunction or leak occur. CO_2 fumigation allows for a great deal of flexibility in

operating methodology, while still being effective. Concentrations of gas over 60% are easy to attain, and easy to maintain. Temperatures above 20°C are likewise easily attainable. The fact that professional fumigators use these systems, and transport them frequently, attests to the fact that they are durable and easy to use.

In light of our experience, and the positive experiences of other institutions, it is surprising that CO_2 fumigation is not more widely used.

REFERENCES

Newton J, 'Carbon dioxide fumigation in a heated portable enclosure', *pre-print of poster presentation at 5th International Conference on Stored Product Protection,* 1990, Bordeaux, France.

Selwitz C, Maekawa S, (Editors) *Inert Gases in the Control of Museum Insect Pests,* The Getty Conservation Institute, 1998, USA.

MATERIALS AND EQUIPMENT

B & G mini fumigation bubble
 Rentokil
 Rentokil House
 Garland Road
 East Grinstead
 West Sussex RH19 1DY
 United Kingdom

Gas Regulators: Harris two-stage regulator
 Harris Calorific Division
 Lincoln Electric Co.
 2345 Murphy Blvd.
 Gainesville, GA 30504
 USA

Gas Monitor: AMC-1011 Hazardous gas monitor, AMC-360 Series transmitter and sensor
 Armstrong Monitoring Corp.
 215 Colonnade Rd.
 South Nepean
 Ontario K2E 7K3
 Canada

BIOGRAPHY

Sue Warren graduated from the Master of Art Conservation programme at Queen's University in 1987. Since then, she has been employed as a conservator at the Canada Science and Technology Museum in Ottawa, working with a wide range of materials and social history objects. Her area of special interest is the horse-drawn vehicle collection, which has necessitated gaining extensive experience with pest management and on-site fumigation. Sue has published two articles on the conservation of carriages and painted floorcloth and presented talks on the subject, in both Canada and the United States. She has also presented two papers on fumigation and pest management.

CHAPTER SIXTEEN

APPLICATION OF CARBON DIOXIDE FOR PEST CONTROL OF BUILDINGS AND LARGE OBJECTS

Gerhard Binker

Binker Materialschutz GmbH, Westendstr. 3, 91207 Lauf, Germany
Tel: +49 9123 9982 0   Fax: +49 9123 9982 22   e-mail: gbinker@binker.de

ABSTRACT

For some years carbon dioxide has been used for the treatment of movable artefacts. The pests are killed by the carbon dioxide and the loss of water. Movable single artefacts can be disinfested in gas-tight, mobile chambers or bubbles. Tests with carbon dioxide also showed that both church interiors and whole buildings can be fumigated using special sealing techniques.

KEYWORDS

Pest control, gas fumigation, carbon dioxide, building

INTRODUCTION

Inert gases, for example, nitrogen, carbon dioxide, noble gases (argon, etc.), are low-reactive components of the air. Insect pests require oxygen and produce carbon dioxide. Manipulation of the levels of these gases or changes in the natural composition of breathing air (78 vol.% N_2, 21 vol.% O_2, 0.03 vol.% CO_2 and the remainder of noble gases, including 0.9 vol.% Ar) have a negative effect on the metabolism of the pests. Very drastic changes are fatal and can kill pests.

Fundamentally, from an economical viewpoint, there are three possible strategies for an inert gas fumigation to eradicate insect pests:

- *Nitrogen fumigation*. Increase the nitrogen concentration to more than 99 vol.% and thus reduce the oxygen content below 0.5 vol.% (controlled atmospheres).
- *Carbon dioxide fumigation*. Increase the carbon dioxide level to over 60 vol.% (modified atmospheres).
- *Inert gas mixture fumigation*. Simultaneously increase the N_2 and CO_2 concentration while forming a low-oxygen atmosphere.

During nitrogen fumigation, the pests die from loss of water rather than asphyxia. In carbon dioxide fumigation there is also a hyperacidity of the insect's blood, which leads to blocking of the nicotinamide-adeninedinucleotides (NAD) due to an excess of protons. Thus carbon dioxide kills insect pests quicker than nitrogen or argon.

Nitrogen and noble gases do not react with artefacts, whereas, in some circumstances, carbon dioxide may react, depending on the concentration, temperature and relative humidity. Carbon dioxide fumigations follow the schedule:

- *Start phase*. The infested artefacts are moved into an airtight container, chamber, bubble or confinement.
- *Flushing or purging phase*. The air still present in these enclosures at the beginning of the fumigation is diluted by introducing inert gas(es) until the desired low oxygen level has been attained.
- *Exposure phase*. The pests are killed, maintaining a low oxygen concentration for a predetermined length of time and periodically topping up the inert gas.
- *Aeration phase*. The enclosures are aerated after exposure and the artefacts are removed.

METHODS AND MATERIALS

During inert gas fumigations, temperature dependent diffusion and permeation processes occur through the interface seals of the enclosures or the foil walls (see Fick's law of diffusion; Wedler, 1987) due to the high concentration gradients. The inert gases diffuse out of the enclosure depending on the temperature, pressure and concentration gradient. At the same time, the oxygen from the ambient air outside diffuses in the opposite direction. This will continue until the normal atmospheric composition of gases is reached again in the enclosure (concentration gradient = 0). Special foils with low gas permeability, e.g. polyvinylidendichloride, are indispensable as a barrier material. In addition, a positive pressure (of at least 5 Pa) of the inert gas has to be maintained inside the enclosure in order to prevent oxygen ingress. As the pests have to be exposed to the modified low-oxygen atmosphere (exposure time depending on the temperature and the humidity) for several days, or possibly even weeks, until they are killed, large amounts of inert gases are required. The amount of inert gas required depends on the volume, temperature, positive pressure, foil material and leakage rate (gas-tightness) of the enclosure.

During the flush and exposure phases, the inert gases are taken from cylinders, containers or tanks depending on the amount required for the treatment. Liquid nitrogen or carbon dioxide from thermally insulated containers is very cold. Care must be taken that the cooling does not damage the artefacts when the inert gas streams into the enclosure. Efficient evaporation devices (heat exchangers) have to be connected between the gas container and the enclosure in order to evaporate the liquid gases and to heat the gases to the desired temperature. At the beginning of the fumigation, the air in the enclosure contains a certain amount of water, measured as relative humidity. If the completely dry inert gas displaces the humid atmosphere in the enclosure, the relative humidity will drop sharply. After a delay, this dry atmosphere would dehydrate artefacts, particularly those made of wood. This can cause cracking and blistering of polychrome surfaces. In addition, oil paintings could become brittle and tear, thus causing irreversible damage. A gas humidifier therefore has to be used to keep the humidity at the starting level. The remaining oxygen concentration is monitored to make sure it is at a level low enough to eradicate the insect pests. Therefore it is essential that the oxygen monitoring devices be calibrated. During the treatment the temperature, humidity and oxygen concentration should be plotted or recorded.

Carbon dioxide fumigation can be carried out in churches and museums at normal temperatures from 16–25°C within 2–6 weeks. An artificial increase of the temperature is not necessary. Therefore, carbon dioxide fumigation is one of the most material-friendly methods of pest control. The fumigation is carried out in gas-tight steel containers, stationary or mobile fumigation chambers, portable foil covers (bubbles) or in foil chambers which can be dismantled. The former are very gas-tight but are very expensive to buy and maintain and are not usually portable. Bubbles can be easily evacuated and flushed with inert gases, but are less gas-tight. As the covers are flexible, there is danger that the bubble materials cling to objects and break off fragile or filigree parts. This is a risk, particularly with heavy, sturdy or textile-reinforced plastic covers. Bubbles have the advantage that they can be folded, are readily portable and can be used almost anywhere. However, there are few bubble models that are sufficiently gas-tight for nitrogen fumigation. A practically gas-tight bubble system with temperature and humidity control as well as oxygen control has been tested for N_2 fumigations since 1988 and has been in regular use since 1993 (Binker, 1993).

Since bubbles are limited in size, detachable gas-tight foil chambers were developed where the volumes can be extended by adding chamber elements. They are suited for both N_2 and CO_2. All moveable museum exhibits or archives can be disinfested with these 'Altarion Nitrogeno chambers'. Temperature, humidity and oxygen concentrations are automatically controlled and recorded,

and there is no risk of damage to the artefacts. The air outside the bubbles or chambers should be monitored to remove danger of suffocation in the event of leakage or during the aeration phase. These fumigation systems can also be safely used in restoration studios where work must continue simultaneous to the fumigation. Fixed cultural property, for instance, high altars and side altars, pews, pulpits, etc., can be fumigated on-site with CO_2 and N_2. For this purpose, they have to be covered with impermeable sheeting that is sealed against the wall or floor sections and made as gas-tight as possible. However, during 'partial fumigations' or 'compartment fumigations' in a church, it is vital that the rest of the structure is not infested by woodboring Anobiids. As fumigations are not preventive, the fumigated parts may be infested again by flying Anobiids from the remaining infested interior. For this reason, it is important that a careful and comprehensive examination of all the wooden parts is undertaken prior to partial fumigation, using established pest management procedures (sound emission detectors, etc.). Fumigations with carbon dioxide gases usually last 2–6 weeks depending on the temperature, humidity and target pest.

Since inert fumigations using various chamber systems have become routine, we decided to investigate whether entire church interiors could be successfully fumigated with carbon dioxide. The churches were carefully sealed using various sealing techniques, pressure tested and subsequently flushed with carbon dioxide. The relative humidity and the temperature inside the churches were constantly measured, recorded and maintained by adding water vapour. Thus the sensitive sacral cultural properties were not harmed due to dehydration caused by the dry carbon dioxide flushed in. A measurement and control device determined the actual concentration of gas and automatically added carbon dioxide to replace gas losses during the long exposure time. The large amounts of required gas (of the order of tonnes) made new logistic concepts necessary to maintain the supply.

Truck tanks delivered the carbon dioxide into a storage tank (8 in Figure 1), from where it was released via pipelines. Prior to introducing the gas into the church, the carbon dioxide was evaporated by means of a heat exchanger (9 in Figure 1) and preconditioned to the church's internal room temperature. In order to keep the amount of carbon dioxide as low as possible, the interior volume of the church was reduced by means of air-inflated balloons.

CASE STUDY 1: CHURCH OF SALMDORF

This is a Catholic church located near Munich, where an important wooden religious sculpture, the Salmdorf Pieta, is housed. The church was infested with woodboring beetles, particularly *Anobium punctatum*. They had infested the pews, altars, pulpit, and organ, as well as the Gothic polychromic Pieta. The church, including the steeple, was tarped with a weatherproof tent in order to improve

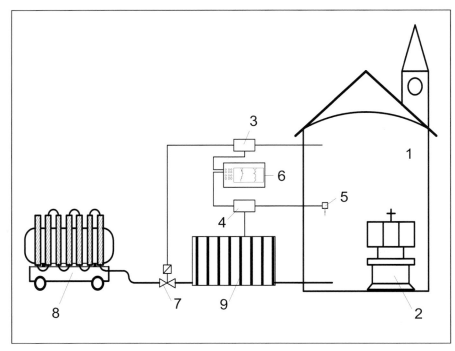

Figure 1 Sketch of a carbon dioxide fumigation of a church

1: sealed church
2: infested artefacts
3: gas analyser for CO_2 or O_2
4: thermo-hygrometer
5: sensor for rel. humidity and temperature
6: data logger/plotter
7: valves (electronically controlled)
8: tank for CO_2 storage
9: heat exchanger

sealing. The tent minimized the gas loss and prevented loss of humidity required for the stability of the wooden artefacts. A moisture and temperature sensor was placed directly on the Pieta to monitor conditions and therefore avoid potential dehydration damage.

In addition to dehydration, there was also concern that the carbon dioxide might chemically change the polychrome of the sculpture. A pigment panel was placed near the Pieta. The pigments were compared with a reference pigment panel after the fumigation. Neither colour changes, nor dehydration cracking, nor damage to the Pieta itself were detected.

In order to monitor the effectiveness of the fumigation, test insects, *Anobium punctatum* and *Hylotrupes bajulus* larvae, were sealed inside wooden blocks. The blocks were placed in the church prior to fumigation. Subsequent examination after splitting the blocks showed that all larvae were killed.

CASE STUDY 2: MONASTERY CHURCH OF SCHAEFTLARN

The church interior was severely infested by *Anobium punctatum*, making fumigation necessary. The church was attached to a monastery containing a private school, which refused to allow fumigation with toxic gases, as an evacuation would have been necessary. As this was not possible, the church authority decided to treat the church with carbon dioxide. The volume to be filled with gas, originally 23,100 m^3 was decreased with air-inflated balloons prior to fumigation. Although this reduced the amount of fumigant that was needed, an enormous quantity of carbon dioxide still had to be maintained at a concentration of about 70% over nearly six weeks of

exposure. Carbon dioxide in liquid form was periodically delivered by tankers and pumped into a large supply tank. The carbon dioxide was automatically withdrawn from the tank and passed through a battery of heat exchangers to convert it to room temperature. A large container heating system regulated the temperature inside the church to 22°C. Sunshine during daylight hours elevated the temperature to a maximum of 29°C. A monitoring and control device measured the concentration of carbon dioxide in the church and automatically topped up the concentration to 70%. This was necessary because leakage occurred during the extended fumigation period. Overall, 1000 tonnes of carbon dioxide were needed to complete the large-scale project successfully.

The relative humidity was kept between 58–70% by additional humidification of the carbon dioxide atmosphere according to the hygrometric chart. Tonnes of evaporated water were injected into the church to prevent objects drying out. No damage was visible after the fumigation.

Bioassay was done using Anobiid larvae and no test insects survived the treatment. Carbon dioxide fumigation of buildings usually presents a low risk of hazardous emissions. Nonetheless, comprehensive safety measures had to be taken. Among them were warning signs at the entrances of the building to prevent unauthorized entry. Emergency medicine and medical equipment including respirators for lack of oxygen treatment were available. The oxygen concentration was monitored to ensure safety when entering fumigated areas, particularly in the basement, where carbon dioxide was more concentrated as it is heavier than air.

CONCLUSIONS

Toxic gases for fumigation of movable artefacts can now be replaced by non-toxic inert gases: nitrogen, carbon dioxide, noble gases, or mixtures of them. The non-toxic inert gases have the advantage that in practical use they do not react chemically with the artefacts, and involve low risk from emission. The death of pests by withdrawing oxygen demands high concentrations of inert gases as well as exposure periods ranging from a few days to several weeks, depending on the room temperature and the moisture content. Since the inert gases from storage tanks or cylinders contain almost no residual moisture, they must be humidified in a controlled manner so as to avoid drying out of the artefacts. Movable single objects (figures, altars, pulpits, furniture, paintings, etc.) including the exhibits of entire museums can be disinfested in gas-tight, mobile chambers or bubbles.

Tests with carbon dioxide showed that both church interiors and whole buildings, including museums and churches, could be fumigated using inert gases following special sealing. The buildings are covered with a form-fitting tarpaulin. The carbon dioxide is supplied from truck tanks. Heat exchangers raise the temperature of the liquid carbon dioxide to the same temperature as the building. The carbon dioxide is humidified and pumped into the building after reducing the volume with gas-filled balloons. The carbon dioxide is maintained at 60 vol.% for 2–6 weeks depending on the temperature in the building. This great technical sophistication makes carbon dioxide fumigations of entire buildings very efficient, although very expensive.

REFERENCES

Binker G, 'New concepts for environment protection and new developments in the fumigation of cultural property', in *Proceedings of the Conference Held by the Restoration Studios of the Bavarian State Conservation Office*, Munich, 22 October, 1993, 90–99.

Wedler G, *Physical Chemistry*, Third Edition, Verlag Chemie, Weinheim, 1987.

MATERIALS AND EQUIPMENT

Binker Materialschutz GmbH (tarps, bubbles, humidifiers)
Westendstr. 3
D-91207 Lauf
Germany
Tel: +49 9123 99820

Linde AG (heat exchangers, carbon dioxide)
Carl-von-Linde-Straße 25
D-85716 Unterschleißheim
Germany
Tel: +49 89 310010

ACKNOWLEDGMENTS

I owe special thanks to my brother Joachim Binker without whose fruitful imagination and tarping skills many of the described fumigations would not have been possible.

BIOGRAPHY

Gerhard Binker, born 6 October 1959, holds a PhD in chemistry. He is the president of Binker Materialschutz GmbH, a company offering pest control in museums/churches and fumigations of artefacts using inert gases and sulfuryl fluoride.

CHAPTER SEVENTEEN

FERAL PIGEONS: A FORGOTTEN PEST?

Susan Rees

Dundee Arts & Heritage, McManus Galleries, Albert Square, Dundee DD1 1DA, Scotland
Tel: +44 1382 432065   Fax: +44 1382 432052   e-mail: sue.rees@dundeecity.gov.uk

ABSTRACT

Pigeon nests attract a whole host of pests that can enter buildings and infest collections, while pigeon droppings not only look unsightly, but are also a health risk and can damage stonework and metal. This potential threat can be overlooked by conservators due to the emphasis on pests within the museum and the fact that care of external buildings is often the remit of other professionals. The author draws from her own experiences as the conservator in a local authority museum, given the task of dealing with a pest infestation caused by nesting pigeons at the McManus Galleries, a 19th century listed museum building. Practical issues, such as investigation, removal of the fouling and steps taken to prevent reinfestation, are covered. The cost of pigeon proofing and various methods currently available are examined, together with the problems of getting money from a museum budget for pigeon proofing and of convincing managers of the seriousness of the threat involved.

KEYWORDS

Pigeons, insect pests, health hazards, pigeon proofing, collections care

INTRODUCTION

The conservator's role

To date, it seems that very little has been written about the problem of pigeons from a conservation point of view. This includes the damage pigeons cause to the outside of buildings and the threat pigeon nests pose to collections by acting as a breeding ground for a host of pests. When discussing the life cycle of museum insect pests, the fact that some originate from bird's nests is often mentioned in texts but taken no further. For this reason, it is important for conservators to look at what is happening to the outside of the building, as well as the inside, in terms of collections care, risk assessment and conservation planning.

It is all very well being able to identify insect pests that threaten our collections but often the source from which they originate is a pest in itself. If we understand more about the feral pigeon and its habits, we may be in a better position to inform and encourage others to assist in implementing preventive measures against these birds.

Most people would surely agree that wild birds enhance the quality of one's life. It is welcoming to hear a chaffinch in a tree outside one's office in the city, trying to compete with the engine noise of a double-decker bus. Birds such as feral pigeons are superbly adapted to living very close to humans, but sometimes this is too close for comfort.

Feral pigeons or *Columba livia* (Figure 1) are descended from the wild European rock dove which was domesticated by humans. They can breed all year round, but are most active during spring and summer. They usually lay two eggs each time, and have been known to raise up to nine broods in a year. As descendants of the rock dove, these birds are well adapted to living on urban buildings, which to them are almost like artificial rock faces. They need very little nest material, often using the building structure itself as the nest base, and can roost and nest on high narrow ledges. They are also very persistent in staying at a site. The ability of feral pigeons to live so close to humans is what causes the problem. They spread disease, contaminate food with their droppings, deface and damage buildings and introduce pests that can pose a direct threat to museum collections.

Figure 1 Feral pigeon (*Columba livia*)

Table 1 Some diseases that can be transmitted to humans via *Columba livia*

Disease	Method of transmission	Symptoms
Histoplasmosis (fungus *Histoplasma capsulatum*)	Inhalation of spores present in soil enriched by faeces (usually of three years age or more) under pigeon roosts	Mild fever, flu-like illness, pneumonia, hepatitis, death
Cryptoccocis (fungus *Cryptococcus neoformans*)	Spores in faecal soil	Cryptoccal meningitis
Ornithosis (psittacosis) (organism *Chlamydia psittaci*)	Spores in faecal soil and feathers. Prevalent in active roosts	Flu-like symptoms, pneumonia, death
Lyme disease (bacteria)	Tick	Fever, muscle ache, arthritis, affects heart and nervous system
Salmonellosis (bacteria)	Ingestion of bacteria in food contaminated with faeces	Headache, fever, nausea, abdominal pain, vomiting

Hazards to human health

People attempting to convey the sort of health hazard they can pose have called feral pigeons 'flying rats'. Like some other bird species, they carry several bacterial infections, called zoonoses, which can be transmitted to humans (Table 1). It is the pigeons' lack of fear of humans and their consequent proximity to us, their habit of living in flocks ranging from twenty to several hundreds, and their eating habits that cause feral pigeons to contract and transmit diseases such as salmonella. Pigeons are classified as a pest bird, along with some gull species, sparrows, starlings and a few others. The Wildlife and Countryside Act (1981) allows them to be legally controlled. Diseases can be transmitted either by direct contact with the body tissue or fluid of the pigeon, such as the droppings, skin particles or airborne spores from faeces (guano). Or indirectly, by ectoparasites such as ticks, fleas, lice and mites that feed on the infected animals and then feed on humans.

Pigeons often roost on the roofs of buildings around heating outlets and air conditioning ducts, which provide warmth. Airborne spores from faeces, insect pests and ectoparasites can travel back down these routes to areas of human occupation. Many of these harmful fungal organisms are present in the atmosphere anyway, but in pigeon guano they exist in very concentrated amounts with which the human immune system is often unable to cope. Pigeon guano contains the fungus *Histoplasma capsulatum*, which can cause histoplasmosis in humans, a respiratory disease manifested by flu-like symptoms. 'Pigeon Fancier's Lung', known as extrinsic allergic alveolitis, caused by inhalation of dust from roosts and feathers, is common among people who keep pigeons and can cause death if untreated. It can occur several years after the pigeons have left (Wilkinson, 2000). The fungus *Cryptococcus* is also found in pigeon droppings and can cause meningitis in people with lowered immune systems,

such as HIV patients (Knight *et al.*, 1993). *Chlamydia psittaci*, a bacterial parasite, can be transferred from birds, such as parrots and pigeons to humans and causes psittacosis or ornithosis (Wreghitt, 1993).

Other diseases, such as salmonella, can be transmitted from the pigeon's guano, via their feet, as they walk over surfaces where human food is consumed or eaten, such as park benches and café tables. *Escherichia coli* bacteria have also been found in pigeon droppings.

Besides diseases, pigeons are host to several species of ectoparasites. Mites and ticks breeding in the droppings and nest debris can cause irritation and rashes on human skin as they bite and suck blood from their hosts. Mites are generally very small, ranging from 0.1 mm to 0.6 mm. Some are just visible to the naked eye, while others require a hand lens or even a microscope to be seen, and so can be difficult to detect before they become a problem. They are generally oval in shape, with no antennae, and range in colour from greyish-black to red. Many mites, including the poultry mite, can survive for 2–3 weeks away from their host, which is plenty of time to wander from the pigeon nest in search of a juicy human. Mites such as the red poultry mite (*Dermanyssus gallinae*) are usually the worst offender in British infestations, causing skin irritation and eczema in humans. The author has had first hand experience of the irritation caused by mites. In this instance, they were coming from a starling's nest, in the wall of a dilapidated prefabricated office building. Mites almost too small to see were dropping down from the wall into an office space; it was not a pleasant experience.

Ticks such as the pigeon tick (*Argas reflexus*) are sometimes a problem to humans. They can survive even longer periods without food than mites, and have sometimes been present 2–3 years after pigeons have been removed from a site (Mourier *et al.*, 1977). Ticks are generally larger than mites and usually easier to detect before they become a problem.

Hazards to the fabric of the building

Most of us are aware of the importance of washing bird droppings off cars as soon as possible because of the damage it can do to the paintwork. The uric acid in droppings is also very corrosive to building materials, such as stone and metals, and often leaves stains that are very difficult to remove. Pigeon infestation on a roof can shorten the useful life of the roof by several years. During the refurbishment of Dundee City Council's Natural History Museum, it was found that the steel girders supporting the cupola had been heavily corroded by excrement from pigeons nesting in the space. Nest material can also block gutters, water conduits and down pipes causing flooding.

Hazards to collections

It is well known that the debris from pigeons nests, such as feathers, droppings, dead chicks and the dead birds themselves, are an ideal food source and nesting material for pests such as moths, dermestids and spider beetles (Mourier *et al.,* 1977; Linnie, 1987). These insects can then find their way into the interior fabric of the building and into collections, as they are attracted to similar foodstuffs within the objects, with devastating results (Table 2). In most cases, it is the larvae of these insects feeding that do the damage, with some, such as hide beetles, dermestes, also creating destruction by boring into material to pupate (Busvine, 1966). The main classes of material threatened are textiles, ethnographic collections and natural history collections. All of these contain a high proportion of organic material attractive to these insects: keratin in feather, wool, hair, baleen, horn, skins and leather; cellulose in plant material; and chitin in the exoskeleton of insect collections. Other materials not normally attractive to these insects, such as synthetic textiles and wood, can also be damaged as the larvae bore or chew their way through them to get to their desired food source.

Insects can enter a building in several ways: adults flying through open windows and doors in the summer, and adults and larvae falling down from the roof space or crawling in through heating or air conditioning outlets on the roof. A pigeon nest has been analysed to ascertain the type and number of insects associated with it (Hancock, 1993). The results were:

- 10 brown house moths (*Hofmannophila pseudopretella*)
- 43 golden spider beetles (*Niptus hololeucus*)
- 15 larder beetles (*Dermestes peruvianus*).

This amounts to nearly 70 insects, along with other beetles, mites, spiders and carrion-feeding blowflies. If this is the number of insects found in one nest, it is easy to estimate how many insects are potentially on the loose around the perimeter of a building, from an average-sized flock of about 20 pigeons.

At the McManus Galleries, early in 1990 shortly after taking up the post, the conservator was alerted by gallery staff to a potential problem when an unusual number of insects were being found on the floor at one end of a gallery. The situation was monitored and a reputable pest control firm was contacted to assist in identifying the pest and tracing its entry point into the gallery. The pest was the Australian spider beetle (*Ptinus tectus*) and, because of its known habit of breeding in bird's nests, the roof space above the area in the gallery was investigated. Decorative turrets in this area were found to contain evidence of pigeons, which had been nesting there undisturbed for many years. They had gained access through open mesh gates, which should have been kept closed to prevent their access (Figure 2). The sight was not a pretty one, with piles of pigeon guano over 300 mm thick, feathers, eggs and the corpses of chicks and adults in various states of decay. Dundee City Council Environmental Health Department was given the unenviable task of cleaning out the turrets and spraying them with insecticide to kill any residual insects. After this operation, the incidence of spider beetle occurring in the gallery dropped dramatically and no more were found. The turrets provided access for roof

Table 2 Insects found in pigeon nests, which are a threat to museum collections

Insect	Materials under threat
Australian spider beetle (*Ptinus tectus*)	Textiles, plant specimens, insect collections
Golden spider beetle (*Niptus hololeucus*)	Textiles, plant specimens, insect collections
Larder beetle (*Dermestes lardarius*)	Untanned leather and skins
Hide beetle (*Dermestes maculatus*)	Untanned leather and skins
Varied carpet beetle (*Anthrenus verbasci*)	Skins, plant specimens, insect collections
Two spot carpet beetle (*Attagenus pellio*)	Insects in natural history collections
House moth (*Hofmannophila pseudospretella*)	Soiled or damp textiles
Case bearing clothes moth (*Tinea pellionella*)	Textile, fur, leather
Common or webbing clothes moth (*Tineola bisselliella*)	Textiles, natural history collections, fur, feathers, skin, wool

Figure 2 Build-up of pigeon guano and other material inside turrets

maintenance, so that contractors and personnel could pass through them, often unsupervised. After this clean-up of pigeon guano, a system was introduced for keeping the mesh gates closed when not in use. However, this system has since failed, underlining the need for communication, understanding and commitment by all concerned towards preventive measures. Pigeons have now been able to re-enter these turrets and a large quantity of nest material and guano, amounting to several centimetres thick, has built up again over the years. Once this material has been removed and the area sprayed, a new approach will be taken by fitting springs on the gates, so that they are self-closing. Notices will also be installed on the gates explaining to users why they must be kept shut.

HOW TO DEAL WITH THE PROBLEM: AN INTEGRATED PEST MANAGEMENT APPROACH WILL BE NEEDED

Getting support

Monitoring by using insect traps is a good first step to find out if there is a problem and if it can be attributable to pigeon nests.

Be prepared for the solution to be neither easy nor cheap. It is estimated that approximately £20 million pounds is spent annually in Britain dealing with problems

caused by urban pest birds, mainly the feral pigeon (Clark, 2000). At the McManus Galleries, in 1990, the cleaning of three turrets, spraying of the area and fitting of about 30 m of wire point proofing, cost in the region of £300 to £400. Today, it would probably cost £2000 to £3000.

It is sometimes difficult to show a dramatic cost–benefit outcome of anti-pigeon measures. Often benefits can only be measured over a number of years in terms of qualitative measures, such as reduction of threat to the collections, improvements in hygiene and improvements in health and safety. It is, therefore, very important to:

- Obtain an understanding and commitment from others by harnessing the advice and support of other professionals such as environmental health officers, architects and building managers.
- Prepare a short report on the problem suggesting some solutions.
- Build up evidence. Where are the birds nesting and roosting and why? Are they being encouraged by someone feeding them nearby? How many pigeons are there?

Photographic evidence is useful, since people are often happy to ignore things they cannot see, but a lurid colour photo of piles of pigeon droppings, feathers and carcasses helps to bring the problem into sharper focus. The local environmental health office are often only too happy to advise of the problems and the health hazards that pigeons cause.

PRACTICAL SOLUTIONS: MEASURES THAT CAN BE TAKEN TO DISCOURAGE NESTING BIRDS

Cleaning up

Roosting and nesting sites usually have to be cleaned out first, but this is not a job for amateurs. This is a dangerous task as spore-laden dust from guano, containing several potential disease organisms, could contaminate the rest of the building or other personnel, causing illness. Even limited contact with such dust has been shown to make people very ill (Dean *et al.*, 1978). Human health could be at greater risk by a clean-up operation that is not done properly, than if the mess were left undisturbed. The clean-up and containment of the guano should be borne in mind when choosing a contractor. Find out how they propose to proceed with the operation. People in the vicinity must be protected from spore inhalation, and spore dispersal must be as minimal as possible. In the United States it is recommended that operators wear full-face respirators with high efficiency particulate arrestant (HEPA) filters and disposable protective clothing, which must be disposed of as infectious waste (USA EHA, 1992). The local environmental health office can usually provide advice on what measures contractors should take and may even be able to supply the names of reputable companies. Ensure the area is sprayed with an insecticide to kill any residual insects, which otherwise will crawl or fly away looking for another host. This could be the museum collections or even nearby humans.

Proofing

Proofing is costly and many authorities try a combination of other methods, such as removing sources of food, appealing to the public and businesses, and scare tactics to discourage the birds.

It is advisable that an entire building is proofed as pigeons are extremely persistent, and if prevented from roosting on one part of the building will move to other parts of the building. In our case, funds were available only to proof the half of the building where the pigeons were currently roosting and nesting and, sure enough, the pigeons merely moved to the other side of the building.

Physical barriers

There are several different sorts of physical barriers on the market and the technology is improving all the time making them more durable and unobtrusive (Figure 3).

Most require some sort of fixing to the fabric of the building, using stainless steel bolts or screws, or the application of an adhesive. Expert advice from architects or engineers on whether this will compromise other aspects of the building's performance should be sought. If the building is listed, the relevant authorities must be consulted.

Sticky gels have been used extensively but have a limited lifespan and leave a black unsightly deposit on stonework that is hard to remove. One modern gel consists of a polybutyl gel sealed with a special fluid (Network Pest Control, 2000). The surface remains as a soft and unstable surface for the birds to walk on, which they do not like. Gel is probably the cheapest option but only lasts for a couple of years before it needs to be replenished. Metal point systems, consisting of stainless steel wires on a UV light-stabilized plastic base, have to be fixed to the building surface, usually using an adhesive such as a natural curing silicone. The points are particularly useful for ledges, parapets, pipes and other areas where netting would not be appropriate. The spikes prevent the birds from landing and are not visually obtrusive from street-level. There are different systems for heavy, medium or light pressure from birds and it is important to choose the right one. The point system used at McManus Galleries ten years ago was easily damaged, the spikes being bent over either by humans or

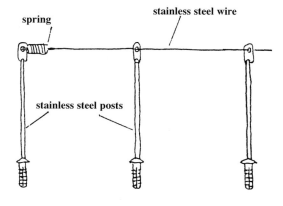

Wire system

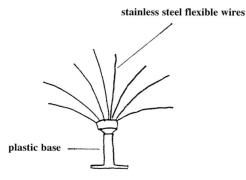

Flexible wires

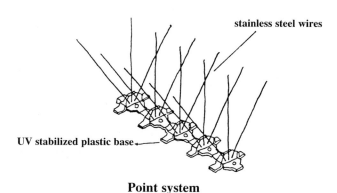

Point system

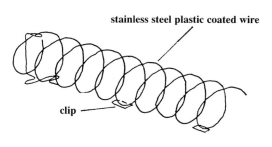

Coil system

Figure 3 Some examples of physical barriers currently available

larger birds. In some cases the adhesive had failed and sections had fallen off. Much of it was damaged by contractors putting in architectural lighting only a year or so after installation. The technology of the adhesive and the points has developed since this installation but, clearly, once a preventive measure has been put in place it has to be checked and maintained regularly and replaced if necessary.

Netting in UV and rot-resistant polyethylene or polypropylene is another frequently used option. Nets are now made in a variety of colours to match the buildings, helping to make them even less visible to the human eye from the ground. Most nets are supposed to last about ten years and can be used for large and small areas. Large nets usually have to be installed by a skilled operator, using a steel or plastic suspension system. The mesh size must be small enough to exclude smaller birds, such as sparrows, if they are also a potential pest. It has been known for pigeons to be excluded, only to be replaced by sparrows, which are able to get through the mesh and move in, undisturbed by bigger birds.

Wire and post systems consist of one or more lines of stainless steel wire, usually plastic coated, tensioned with springs a few inches above the surface to be protected and fixed into the structure with nylon or stainless steel fittings. They are recommended for light use only.

Another system consists of stainless steel wires on a plastic cone, slotting into a plastic base, which is glued or screwed to the building structure. The wires are flexible and wave about in the wind, thus disturbing the pigeons. The ends of the wire are plastic-tipped to prevent damage to the birds. This system is useful in areas where access is required for maintenance as it can be removed and then replaced once the maintenance is finished.

Metal coils consist of a stainless steel coil held in place by a variety of fixings. These are also easy to remove and replace from their fixings when access is required. One problem that has been known to occur with these removable systems, is that sympathetic misguided people remove them deliberately from areas such as from underneath office windows in order to allow pigeons access to where they can be fed.

Scare tactics

Menacing-looking plastic models of birds of prey are marketed as being able to scare pigeons away. In most cases, it has been found that this method only works for a short period of time before the pigeons realize that there is no threat from the model and ignore it. Natural predators of the feral pigeon in Britain are the female sparrowhawk, goshawks and the peregrine falcon, but there are rarely enough of them in one area to make a significant impact on a feral pigeon population. In some areas, birds of prey have been used deliberately to keep the pigeon population down, by employing a falconer. It has been found that only regular visits to a site using this method are effective, such

as at a London football stadium where the falconer visits every couple of weeks, in order to keep up the pressure on the pigeons to stay away.

PREVENTIVE MEASURES

To be successful, prevention will usually involve several approaches in combination.

Education and information

In order to gain support from the public and colleagues, information is important. As shown in our own situation, gallery staff and other colleagues in the museum are vital in alerting those concerned with insect monitoring to potential threats. The conservator held a seminar for all museum staff, from gallery attendants to curators, on insects pests and the threat they pose to the collections. Real specimens were passed round, and a simplified explanation of the life cycle and habits of the insects was given, with some graphic examples of the destruction they can cause. Staff were issued with identity sheets to post in their offices bearing enlarged illustrations of the most common pests and entitled 'public enemies'. This was a few years before the excellent colour poster from English Heritage and MGC was produced (English Heritage, 1999).

A system was also implemented by the conservator whereby attendant staff were asked to collect insects they found in the galleries and stores in the course of their duties. A stock of small glass specimen tubes was provided, each carrying a label to record who found the specimen and where and when it was found (Figure 4). This exercise has proved very useful in getting participation and interest from staff, giving them a feeling that they are helping with preventive conservation and the crusade against insect pests. It is also vital in adding to our picture of which insect pest is present and where and when it is occurring. With inevitable changes in museum personnel, training needs to be repeated every few years.

Building maintenance

Buildings are intended to protect people and collections and not to act as a conduit for their destruction, which is

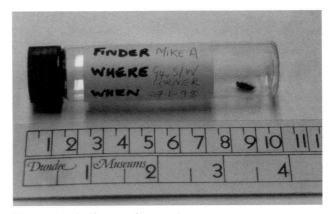

Figure 4 Tube for recording specimens

what a building can become if it is not maintained properly. One of the problems of pigeon infestation is that, because they roost out of view from most people and in usually inaccessible areas, it can be several years before they are noticed as a problem. By this time, it is likely that several inches of guano and other debris will have built up in various areas around the building, and the incursion of insects into the interior could be well under way.

The conservator should try to work closely with the building manager and to persuade them of the problems that pigeons pose and what can be done to discourage them. One of the best ways of discouraging birds from nesting is building maintenance. Regular checking and cleaning of roof areas, downpipes and guttering for nests and debris will help to discourage roosting and probably prevent the inevitable flooding from a blocked water conduit. Air conditioning and heating ducts also need to be maintained so that they do not become damaged in any way that might provide an entrance route for insects into the building.

Contractors working on a building also have to be closely supervised. The author has found that if pigeon proofing gets in their way, it is often damaged and the proofing system left severely compromised. Easily removable types of proofing, now available, would be very useful for this type of situation. Cases have also been reported of a contractor inadvertently sealing a pigeon into an air-conditioning duct. The pigeon subsequently died and it led to a collection infestation (Jeffers, 2000). Making sure that food sources are not available, keeping litter swept up, litter bins emptied and refuse bins closed, are all important to deny the birds of a nearby food source. Noticeable amounts of pigeon droppings on the exterior of a building look unsightly, giving the appearance of an uncared-for property, and can often be an indicator of more deep-seated problems within, such as lack of resources and lack of commitment to collections care.

Public awareness programmes

Pigeons in cities rely on people feeding them for their survival and growth in numbers.

One of the problems encountered in curbing pigeons seems to be some of the public's attitude towards them and to the people trying to control their numbers. The press often have a field day with amusing headlines such as 'Pooping pigeons pose putrid problem' (Kingman, 1988) and 'Pigeons: Not a problem to poo-poo' (BBC News Online UK, 1999). The authorities, local council and environmental health officers are portrayed as the baddies, and the pigeon-loving people as the goodies. Banning feeding of pigeons can upset people but if they were made more aware of the hazards to human health and building fabric that these birds pose, then the task of curbing them would be easier. In fact, many local authorities are trying more humane ways of discouraging pigeons, instead of shooting or trapping them, as they have found that other

Figure 5 Public notice requesting no feeding

pigeons merely move in to fill the gap left by those that have been removed (Haag-Wackernagel, 1999). Many are now appealing to public and local businesses through awareness campaigns using leaflets and posters, informing them of the diseases pigeons harbour and asking them not to provide food for the pigeons, deliberately or otherwise (Figure 5). Denying the pigeons sources of food forces them to move to another site and also keeps the number of pigeons down by natural competition. Removing sources of food by ensuring proper refuse disposal and discouraging feeding by public awareness programmes has been found by some as the only successful long-term method to keep the numbers of feral pigeons at manageable levels (Haag-Wackernagel, 1999).

Planning

It is important, having dealt with the immediate problem, not to forget the ever-present threat of pigeon damage. Pigeons are very persistent and superb opportunists, and will resume roosting and nesting in an area if barriers fail and allow them access. Physical pigeon barriers will need to be inspected at least annually, and probably some of them will have to be replaced every few years. It is important to try and persuade managers to set aside money every year to fund yearly inspection, replacement and repair. Some pest control companies and local authority environmental health organizations offer a maintenance contract whereby, for a fee, they will inspect buildings regularly to check on proofing and ensure no reinfestation has occurred. Such measures will avoid the necessity of having to spend huge sums of money replacing large areas of unmaintained pigeon proofing, or rectifying associated damage and infestation every ten years or so. If a museum has a collections care policy or a conservation policy, the conservator should endeavour to get measures for inspection and proofing included as part of the preventive conservation programme, and this should certainly be included in any integrated pest management (IPM) programme.

CONCLUSION

A conservator's interest in pest control should extend to the exterior of a building, as well as the interior. Taking steps against bird infestation of buildings in which collections are housed, should be part of a preventive conservation plan and IPM by awareness, planning and budgeting,

The conservator cannot be expected to tackle this problem alone and will need to enlist the help of other professionals. Pigeons can be successfully discouraged from buildings using a combination of low cost methods, namely, maintaining the fabric of the building, keeping standards of waste disposal high and education of colleagues and the general public. In many cases these measures will also have to be backed up by more costly proofing.

REFERENCES

BBC News Online: UK, 'Pigeons: Not a problem to poo-poo', January 19, 1999, http://news.bbc.co.uk/hi/english/uk/newsid_257000/257284.stm

Busvine J R, *Insects and Hygiene,* Methuen, 1966, London.

Clark B, 'The cost benefits of bird management', in *Urban Bird Management: An Evaluation*, 12–13 Pest-Ventures 2000.

Dean A, Bates J, Sorrels C, Sorrels T, Germany W, Ajello L, Kaufman L, McGrew C, Filts A, 'An outbreak of histoplasmosis at an Arkansas court house with five cases of probable re-infection', in *American Journal of Epidemeology,* 1978, **108**(1), 36–46.

English Heritage, 'Insect pests found in historic houses and museums', poster, English Heritage, Museums & Galleries Commission, 1999.

Haag-Wackernagel D, 'Wildlife management of the feral pigeon *Columba livia*', 1999, http://www.unibas.ch/dbmw/medbiol/haag _1.html

Hancock G, 'Museum pests from pigeon nests', in *Journal of Biological Curation,* 1993, **1**(3/4), 41–43.

HMSO, *Wildlife and Countryside Act,* 1981, Her Majesty's Stationery Office, London.

Jeffers W, 'Light traps', Conservation DistList Instance, 2000 14:32.

Kingman A, 'Pooping pigeons pose putrid problem', in *Las Vegas Review Journal,* July 17, 1998, http://www.lvrj.com/lvrj_home/1998/Jul-17-Fri-1988/news/7863467.html

Knight F R, Makenzie D W, Evans B G, Porter K, Barrett N J, White G C, 'Increasing incidence of cryptococcosis in the UK', in *Journal of Inflection*, 1993, **27**, 185–191.

Linnie M J, 'Pest Control: a survey of natural history museums in Great Britain and Ireland', in *International Journal of Museum Management and Curatorship*, 1987, **6**, 277–290.

Mourier H, Winding O, Sunesen E, *Collins Guide to Wildlife in House and Home*, Collins, 1977.

Network Pest Control 2000, http://www.network-pest.co.uk.

USA EHA, *'Managing health hazards associated with bird and bat excrement'*, United States Army Environmental Health Agency, USA, 1992, TG No. 142.

Wilkinson P, 'Disease alert over pigeons after death', Times Newspapers Ltd, 2000, http://www.times-archive.co.uk/news/pages/tim/2000/05/26/timnwsnws01020.html

Wreghitt T, 'Chlamydial infection of the respiratory tract', in *Communicable Disease Report Review,* 1993, **3**(9), 119–124.

ACKNOWLEDGEMENTS

The author would like to acknowledge support from The Conservation Bureau, Historic Scotland and Dundee City Council to enable participation in this conference. The author would also like to thank Brian Gilmour, Pest Control Unit, Tim Fourie, City Centre Building Manager and Jo Sage, Dundee Arts and Heritage, all of Dundee City Council, and Liz and Andrew Robertson Rose.

BIOGRAPHY

Susan Rees holds a degree in Archaeology from the University of Cape Town. She graduated in Archaeological Conservation from University College, Cardiff in 1981. She has worked for English Heritage and York Archaeological Trust, and has been employed by Dundee City Council since November 1989 as Conservator for social history, archaeology, ethnography and decorative arts collections at Dundee Arts & Heritage. As the only conservator in the museum, her remit also includes environmental monitoring, pest monitoring and advice on the environment for the stores, galleries and for exhibitions of all the museums collections. She is an accredited conservator and a committee member of the Scottish Society for Conservation and Restoration.

CHAPTER EIGHTEEN

PRINCIPLES OF HEAT DISINFESTATION

Thomas J K Strang

Canadian Conservation Institute of Canadian Heritage, 1030 Innes Road, Ottawa, Ontario, Canada
e-mail: tom_strang@pch.gc.ca

ABSTRACT

Heating is an effective way to eliminate insect pests from collections. Relatively short exposures are needed to assure efficacy, so the utility of heat is relatively quick compared to other methods, and the universal susceptibility of insects. The ability to scale from small to large volumes of affected objects, and availability worldwide, largely without encumbrance by patent, royalty, and regulatory approval, also make heat a valuable addition to the tools we require to prevent loss of cultural property. The deleterious effects of heat treatment are quantified so relative risk can be discussed. The magnitude of protection conferred by a vapour resistant enclosure during treatment is illustrated along with examination of mould risk. Heat disinfestation has been used for centuries, both within and outside the context of cultural property, and remains a valuable tool, augmenting our ability to save property from loss to insect pests.

KEYWORDS

Heat, thermal, insect, disinfestation, collections, enclosure, ageing, mould, risk

INTRODUCTION

The choice of thermal disinfestation over alternatives is best made if its principles are well understood. This paper reviews key concepts from which the argument for heat disinfestation is built. Sections examine the effects on the target organisms and the target objects when viewed as both material and assemblage. Much of this summary is based on the author's previous publications on thermal control and extended to explanations and examples derived since that time.

To date, heat has been mostly employed when alternatives are deprecated, either due to efficacy, cost, timeliness, availability, or scale. Part of this deprecation comes from due caution fostered in the conservation profession as the course of minimal intervention. However, the argument for use of heat requires understanding of concurrent events within the atmosphere and materials being treated, which are not generally taught. This paper is meant to assist in following this argument, and show how the delivery method described within greatly reduces the magnitude of the potential effects on objects, compared to the commonly perceived threat.

FIRST PRINCIPLES

This section presents physical concepts, which are fundamental to understanding heat disinfestation. An attempt has been made to place them in step-wise fashion towards comprehending the dynamics of heating objects within vapour resistant enclosures for the purpose of killing insect pests. Some basic comprehension of scientific principles is presumed, but key areas are dealt with at length.

Partial pressure

The total pressure of any atmosphere is the sum of all pressures contributed by its gas and vapour components. The contribution of each component is termed a partial pressure. Partial pressure is directly related to temperature. If you raise the temperature of water in an enclosure, the partial pressure of water vapour increases and if you lower the temperature, the partial pressure drops.

Near absolute zero (−273.15°C), vapour pressure is practically non-existent and matter is essentially in a solid state. At boiling point, the partial pressure of a substance (saturation vapour pressure) is the same as the surrounding atmospheric pressure, and given the right conditions of leakage could fully replace the atmosphere in a restrictive container through dilution (Figure 1). Saturation vapour pressure is a limit. The atmosphere over a bowl of water will not support a partial pressure of water greater than x, where x is the saturation vapour pressure for water at the temperature of the system. Any excess vapour will condense.

An atmosphere can have component vapours at less than their saturation pressure, depending on availability. The quantity of vapour present in the system is moderated by events other than temperature. Temperature simply sets the maximum for the portion of gas present. This relationship for water is fully illustrated in a psychrometric table (Anon. and Mackintosh, 1985).

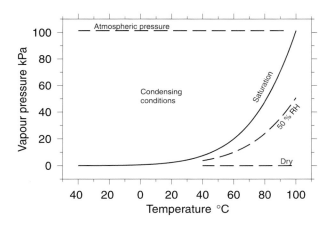

Figure 1 Saturation vapour pressure of water against temperature. Data is from Weast (1972–3)

Humidity may be qualified as relative or absolute humidity. Absolute humidity is the mass of water vapour per unit mass of dry air. Relative humidity (RH) is the ratio of the measured vapour (partial) pressure of water to its saturation vapour pressure at a known temperature. A value of 0.5 RH or 50% RH means that there is only half as much water vapour measured as pressure in an atmosphere as there could be for a given temperature (Figure 1).

In the simple case when water vapour is at a higher partial pressure in one part of a system, water vapour will migrate to the area of the system with the lower partial pressure until equilibrium is reached. To get there, it will migrate by convection, diffusion, or permeation.

If an area of lower temperature exists within a system, convective mixing of air results in an averaging of air temperature, but lowered saturation vapour pressure for the cooled air and hence higher RH.

When a system contains organic matter and a temperature gradient the dynamic is less simple. Vapour permeates through a material barrier between two voids (wall), and vapour within the material equilibrates with the outside world. One way to calculate the potential for motion is by multiplying the RHs (expressed as a fraction) by the associated saturation vapour pressure of water, to get the partial pressures in regions of differing temperature (Latta and Beach, 1964). When these pressures differ, water vapour moves.

In an open system, a portion of an organic barrier can become saturated by permeating water vapour if its temperature prescribes an internal saturation vapour pressure lower than the partial pressure of water in the air supply. Further vapour movement into the barrier condenses until the material saturates. This is a mechanism that damages buildings, and is combated by properly installing a vapour barrier to choke off the supply of water before it enters the wall. Other mechanisms are described in Michalski (1996).

The other available control is insulating or otherwise preventing temperature differences within a system, so that the saturation vapour pressure of the water remains higher than its partial pressure throughout the system (Latta and Beach, 1964). With the exception of insulating heat chambers to save on fuel cost, or backing a solar bag, we need to encourage unobstructed heat flow within the system to effect safe treatment. We use vapour barriers extensively in thermal pest control to resist moisture flow.

Summary: Temperature limits the maximum amount of water vapour that can be in any atmosphere. This limit is unaffected by any other gaseous component. Differing partial pressure (representing concentration gradient) provides the motive force for moisture movement in materials undergoing thermal change. Sufficient temperature change or differential in a system can result in condensation of water vapour, but only when there is a sufficient supply of water. Vapour barriers are used to significantly reduce the rate of moisture movement by eliminating convection and diffusion flow, and slowing permeation.

Moisture content

The previous section ended with moisture concerns. Moisture is used in the sense of water that associates intimately into the molecular structure of materials and subsequently fills capillaries. The standard concept is one of the water first grouping at charged sites, and as more water associates, it layers on top of these initial water molecules and penetrates holes between polymer molecules, ending with water filling pores, or solution of the substance (van den Berg and Bruin, 1981).

Moisture content (MC) on a dry basis is calculated as the ratio of mass of substance containing water minus its dry weight, divided by the dry weight of the substance. Given as a percentage, MC is the weight proportion due to water content. One-inch lumber stored outdoors undercover in a temperate climate will vary by 3% to 5% annually. Depending on the species, the MC will vary from a minimum of 11% to a maximum of 18%. Wood that is stored indoors has a similar annual cycle, but ranges between 6% to 14% depending on species (Baxter *et al.*, 1951).

Unlike vapour and humidity, saturation is more difficult to define with moisture content. It is generally a matter of definition either when the free space around molecules in a solid is filled with water, or when all large-scale pores are also filled. Dryness is often defined by lack of mass change under vacuum, or heating.

Equilibrium is another way of saying wait a long time for all measurable change to stop. The equilibrium moisture content (EMC) of a material is in balance with the equilibrium relative humidity (ERH) in the atmosphere bathing the solid.

Volume change of organic matter correlates with moisture content associated with molecular structures, or thermal effects. When the volume change is unequal in directions relative to the orientation of molecules or cellular structure, the response is called anisotropic. Wood is an example of anisotropy, as most changes in dimension occur in the tangential direction to the grain rather than along it.

Summary: Organic solids absorb and release significant amounts of associated water in response to the water vapour present in the surrounding atmosphere. The dimensions of organic solids are altered by changes in moisture content proportional to the volume of water transferred, but often expressed in an anisotropic (asymmetrical) manner.

Sorption isotherm

Sorption isotherms were developed to quantify the relationship between atmospheric moisture surrounding a substance, and the moisture contained within the substance. Isotherm means the sorption is measured at a constant temperature. Determination of a sorption isotherm uses an essentially infinite source of conditioned atmosphere relative to a rather small sample, to facilitate equilibrium. A full description of numerous methods is given in Gál (1975).

The sorption isotherm is affected by the moisture history of a material. If we start with wet material and place it into a series of drying atmospheres, we can measure the desorption or the loss of water toward the equilibrium state. If we start with a dry material, we can measure adsorption in moist atmospheres.

The equilibrium states of these two approaches are commonly not identical. Desorption isotherms typically have higher EMCs than the adsorption isotherm. This property is called hysteresis – the separation of two isotherms for the same material at the same temperature. The hysteresis in cotton can be 1–2% for a region of 20–80% RH. Two blocks of material brought to equilibrium in the same humidity, one by adsorption and one by desorption, can sit nearby for years and not drift to a common moisture content because there is no difference in vapour pressure exhibited over the two samples (Urquhart, 1960).

In practice, when adjusting to humidity changes in the middle range, the isotherm of a material mimics the general shape of the outside curves, but values are somewhere within the laboratory curves (Urquhart, 1960). Published curves are an outer boundary, and the current EMC of a material depends on its history of exposure. We will prevent large EMC shifts during heat disinfestation so hysteresis can be safely ignored.

When sorption isotherms for a material are compared for several temperatures they show a trend (Figure 2). The higher temperature isotherm exhibits lower moisture content for the same RH. Simply stated, more heat in the system creates increased molecular motion and the moisture increasingly favours the vapour phase over sorption onto solids. As the temperature rises, a higher vapour pressure of water is needed to hold the moisture

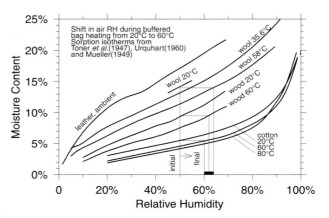

Figure 2 Humidity change with temperature in enclosures. Data is from Toner *et al.* (1947), Urquhart (1960), Mueller (1949)

content steady. This will be accomplished naturally by enclosure. The 80°C cotton curve shows a crossover effect noted in other materials (Urquhart, 1960). This convergent trend is helpful in limiting a rise in RH in treatments that run too hot.

Summary: Sorption isotherms are experimental extremes to plot boundaries of moisture content against RH. The equilibrium moisture content depends on the history of exposure to humidity change, but sorption isotherms give us boundaries of the possible values. As EMC changes are limited, hysteresis can be ignored. The rise of RH in an enclosure is roughly predicted from a family of sorption isotherms.

Enclosure and humidity buffering for dimensional stability

The sorption isotherm represents EMC imposed by ERH in infinite supply with unrestrained time and at constant temperature. Thermal control inverts this so that RH imposes EMC by enclosures, which create conditions of restrained vapour supply. The remaining differences, caused by changing temperature for short periods of time, are examined in the next subsection.

Thermal control for pest eradication uses enclosed systems to its advantage. A sealed bag around an object contains a relatively small volume of air compared to the volume of the object, and is largely leak-free, except for a slow permeation of gas and vapour through the bag wall. This creates a situation where the moisture content of the enclosed material influences the humidity within the enclosure, rather than the moisture content being driven by an effectively infinite source of atmospheric water. This effect is commonly called 'humidity buffering'.

For museum displays, Thompson (1964) determined the minimum ratio (buffering capacity) to control RH change from thermal cycling to be 1 kg of wood to 1 m³ of air, in a swing of 20°C from 15°C to 35°C. Adding even more buffer simply reinforced the already minimized RH change and broadens control through a wider temperature swing. The use of the term 'bagged' or 'enclosed' in this paper implies this latter state.

When heated without a vapour barrier, the moisture content of organic material will drop as water desorbs. When heated inside an enclosure (marked 'initial' in Figure 2), a small amount of desorbed water will elevate the RH (partial pressure) to the RH equivalent (marked 'final' in Figure 2) to the original EMC on the high temperature sorption isotherm. Without motive force, moisture transfer stalls and desiccation does not proceed. At 20°C and 50% RH, 1 m³ of air contains 7.4 g of water vapour. At 60°C and 63% RH (example plotted on Figure 2), 1 m³ of air contains 91.4 g of water (Anon. and Mackintosh, 1985). The difference between these concentrations (84 g), is the required contribution of desorbed moisture if the objects in a bag are to buffer the free space and retain a constant EMC.

The density of seasoned Canadian wood species ranges from 250 kg/m³ to 750 kg/m³ (Baxter *et al.*, 1951). At 10% EMC there is potentially 25–75 kg of water available for exchange in 1 m³ of wood. When about half of the volume is air in cell voids, only some 42 g of this water would need to vapourize to rebalance the EMC at 60°C. This represents the situation of tightly bagged timber.

Examples from Figure 2: For a wood object sitting in 1 m³ of air at equilibrium with 50% RH, heated from 20°C to 60°C, the wood needs to contribute 84 g of water to forestall EMC change. One kilogram of wood at 10% EMC (100 g water) would stand to lose about 80% of its water, desiccating the object. Twenty kilograms of wood would lose 4%, similar to maximum annual fluctuations. Eighty kilograms would lose 1% EMC. A drop of 1% EMC is equivalent to a 5–10% reduction in RH of the common materials shown in Figure 3. The ASHRAE guidelines (Anon. and Parsons, 1999) give ± 5% RH as a rating of 'no risk of mechanical damage to most artefacts and paintings' (type AA) and ± 10%RH as a rating of 'small risk of mechanical damage to high vulnerability artefacts' (type A).

From this, the recommended minimum packing for self-buffering heat treatments should be greater than 100

kg/m³, which is still less than half the volume of a bag with the lightest of common woods. As the density of wood can exceed many hundred kg/m³, there is wide scope for safely enclosing various shaped objects with bags while being reasonably conformal. Obviously, minimizing air volume is beneficial through conformal bagging, or adding exchangeable water, by added dunnage in the form of cotton sheet over-wraps. For a full treatment of how buffering can stabilize EMC in tight or leaky structures, see Michalski (1994).

The coefficient of thermal expansion for wood in its tangential (most reactive) direction ranges from 3.9×10^{-5} to 8.1×10^{-5} per °C (USDA, 1987). A 40°C shift in temperature would result in a 0.3% swelling in the tangential direction. This is not enough to cause fracture in constrained glassy polymers such as most paint on wood, as their elastic limits are between 1% and 3% elongation at break (Michalski, 1991). Therefore, in itself, heat used to disinfest is incapable of damaging constrained assemblages through the effect of expansion, but other factors need to be considered.

For bagged artefacts, the total amount of moisture within the bag remains constant and the EMC of components changes very little (Richard, 1991). As a result, thermal expansion dominates in these situations and the resulting physical movement is too low to cause damage to sound artefacts.

However, for unbagged wooden materials, the loss and gain of moisture dominates dimensional change. The moisture coefficient for wood in the tangential direction ranges from 2.0×10^{-3} to 4.5×10^{-3} % EMC change, and 1% EMC equates to 5.9% RH (Richard, 1991). Using median values, a 1% EMC change will be about 60 times the dimensional change caused by 1°C. Using the equation by Richard (1991), which is valid for a wood buffered enclosure with minimal air space:

$$RH_{final} = RH_{initial} + 0.35 \left(T_{final} - T_{initial} \right)$$

If an unbagged wooden object were subjected to a 40°C temperature increase it would experience a 2% decrease in EMC even when kept at the same RH (see Figure 2), causing a 0.9% shrinkage in the tangential direction. This is approaching the elastic limit of glassy polymers (Michalski, 1991), which indicates that caution should be exercised with assemblies such as painted wood. Bagging artefacts before heat treatment circumvents this problem (Strang, 1995).

A beneficial trend comes from the thermal expansion. While some moisture is lost to vapour in the bag at higher temperature and causes minor shrinkage, thermal expansion counteracts this dimensional change. Through the heating/cooling cycle in an enclosure, minor moisture loss is balanced by thermal expansion and further reduces strain on joints.

The vapour barrier of choice is 150 μm (6 mil) low-density polyethylene. This film is either clear or pigmented,

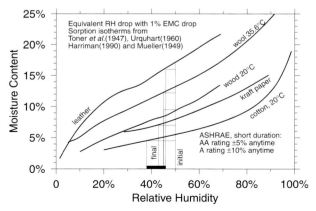

Figure 3 Equivalent RH drop for a 1% EMC drop. Data is from Toner *et al.* (1947), Urquhart (1960), Mueller (1949) and Harriman (1990)

is heat-sealable, widely available, and inexpensive. With a melting point above 100°C and a brittle temperature below −50°C (Andrews and Dawson, 1986), low-density polyethylene is an ideal barrier film for thermal treatments. Moisture permeability increases with heat and is directly proportional to thickness (Figure 4).

While polyethylene can be penetrated by some insects, Highland (1984) rated 125 μm (5 mil) polyethylene as fair, and 250 μm (10 mil) as good for insect resistance. Five-year trials with woollens stored in kraft-polyethylene laminate showed only unsealed materials to be affected by mould and insects (Bry *et al.*, 1972). Artefact bagging can be used with confidence that effort is not being wasted (see below on mould risk).

As folds and seams are the major route of infiltration by penetrating insects (Newton, 1998), seams should ideally be closed with a wide heat seal (1 cm) and trimmed flush to eliminate crevices. Taped seals or rolled and clipped seals will not fare as well in the long term, but may be used. A humidity indicator strip should be dated and placed inside the bag for monitoring long-term storage. Heat disinfestation uses the enclosure to quarantine, retain moisture balance, divide collections into manageable units, and direct heat to where it is needed. The solar bagging application (Strang, 1995; see below) extends the enclosure to become the heat source for carrying out the treatment and protection from environmental disturbance. Polyethylene bags on storage shelves are generally no more combustible than the organic objects they protect. In practice, the Canadian Conservation Institute (CCI) has found that polyethylene enclosures absorb significant amounts of infrared radiation from localized fires, prevents widespread sprinkler and smoke damage, and lowers the overall level of contamination. Even when shrivelled onto objects, the film has often been easily removable.

Summary: The primary function of a vapour resistant enclosure during heat treatment is to restrain moisture loss in the object. With conformal bagging, but not obsessively so, the minor equilibrium moisture content change will result in insignificant dimensional change with little risk of damage. There are currently studies underway to systematically observe any effects on collections being treated. Fast reacting dunnage can reduce drying of thin elements. Thermal expansion counteracts minor loss of moisture during heating. Polyethylene is a good enclosure due to ease of sealing, widespread availability, low cost, long-term stability, and advantages in long-term collection preservation, which offset the labour of bagging.

Rates of moisture and heat equilibration

Restricting dimensional change with vapour barrier enclosures is sufficient to protect objects from damage they would experience without the bag. An examination of the process with time and relative rates of change in heat and moisture content provides information leading to establishing effective treatment schedules.

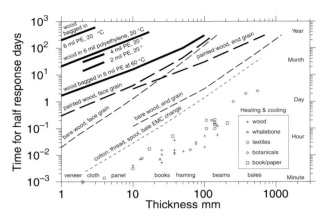

Figure 4 Time for half response of the relative moisture content and heat movement to occur. Data is from Strang, 1995

The curves in Figure 4 represent the time needed for half of the total EMC change to occur. Wood which is bare, painted, and bagged, can be modelled as infinite sheets exposed on two sides (Michalski, 1993b; Crank, 1975), and the cotton data comes from measurements on threads, spools and bales (Crank, 1960).

The time for half hygrometric response was chosen as it coincides with the more rapid and linear portion of the exponential decay curves, which fit equilibration data. To roughly calculate the time for completion, multiply the half response time by a factor of three (88% complete) to five (97% complete).

The examples were chosen to show the resistance of different barriers to moisture movement: diffusion through the boundary layer of air (no bag), permeation through a paint layer, and permeation through a sealed polyethylene bag. Within the range of thickness considered for wood and cotton, the rate of response is primarily determined by the coating or bag. At greater thicknesses (right hand side of Figure 4), the diffusion characteristics of the material inside the coating or bag dominate.

The −20°C and 60°C curves for polyethylene bagged wood illustrate the effect of temperature on barrier permeation. The contribution of film thickness is also illustrated, calculated from the data in Pauly (1989).

The heating of materials is also plotted in Figure 4, where the points represent experimental findings. Wood samples include bagged veneers, structural walls and massive timbers. Books range from a slim volume to full depth on a packed library shelf. Textiles range from carpets to bolts. Botanicals range from a few herbarium sheets, a full cabinet shelf, to bales of tobacco. Cancellous whalebone represents the most insulating form of bone. These examples were extracted from the literature on heat/cold treatment or developed by the author to complete cases similar to the ones encountered in collections.

The significance of the curve and point distribution is that heat (and cold, as it is the reverse flow of the same case) moves faster in and out of materials than moisture at room temperature, even for end grain or fibre exposed to air with no other barrier. However, for very thin objects at elevated temperature, there is a risk of moisture moving almost as fast as heat, so a vapour barrier must be used. An existing paint film would suffice for short treatments of thicker objects, however, a simple inexpensive polyethylene bag is superior and has many advantages over existing surface films at the elevated treatment temperature, and reduces the chance of surface effects.

The overall advantage of bagging objects for heat treatment is a thousand times delay of hygrometric half time relative to the half time for heat (vertical separation of lines and points, Figure 4). The necessary multiplication of exposure time by five to get full penetration of heat is still inconsequential in comparison, even as it approaches moisture response times for the unprotected cotton.

The question of how moisture content changes with mixed materials is of great interest in the food industry (Gál, 1975). The resultant EMC of a finely divided mixture (food) is practically calculated as a mass-proportional mean of the isotherm differences (van den Berg and Bruin, 1981). A parallel concern is safety of multi-material objects with divergent sorption isotherms during a bagged heat treatment. The materials will attempt to independently adjust the EMC, as shown in the multiple final paths in Figure 2. In practice, a new equilibrium RH is struck, based on the proportion of the materials. The relatively parallel and straight middle portion of sorption isotherms mitigates the effect of mixing materials, as the difference in RH between materials is kept small.

As well as surface moisture leaving to perform the buffering effect, moisture will also move within the object. Any difference in temperature through an object will result in a parallel EMC gradient. The major risk to the object is the initial requirement to buffer the enclosed air space, and mitigation of this risk has been described. As the heat flows into the object, desorbing moisture will elevate humidity in cellular voids and fibre gaps towards a new equilibrium vapour pressure. As the external vapour pressure is already high (buffered air in the bag) there is nowhere for moisture to go except towards the cooler core in the direction of heat flow. This moisture would then be adsorbed. However, sorption requires the evolution of heat, which augments the externally provided heating wave. The coupling of simultaneous moisture and heat transfer 'waves' is described in Crank (1975). What is useful for the topic of this paper is that significant amounts of the moisture in motion will not have anywhere to go to cause significant stress to the object during the course of heating and cooling. The voids cannot take on enough vapour. Moisture will in large remain bound to the object, performing its role contributing to dimensional stability.

Once the object is at an even temperature, the moisture gradient should seek equilibrium, but we generally initiate cooling at this point which reverses the gradients, and moisture will now migrate from the enclosed air onto the object's surface and the matter will take on vapour from its

cooling voids. As the movement of moisture in a material is slower when it is dry than when it is moist, the regain will not be symmetrical with time. Some minor hysteresis lowering of EMC may also contribute to the resultant surface EMC, leaving this zone slightly drier than its initial state. This is where dunnage is effective in protecting finer elements from drying.

Any losses, for example permeation through the bag, seal leakage, or dumping condensation upon opening the bag, will also lower the final EMC, but not at a level that causes undue strain.

Condensation forms on cooling when the object is far enough from the bag wall so that the dew point is inside the bag. Heat will flow out of the object and also continue into its centre if a gradient was still in effect when heat was turned off. As the surface cools, it will adsorb moisture from both the air and the warmer core, reversing the trend established during heating.

The only way to reduce this small but unavoidable 'sloshing' effect of moisture movement within the object is to optimize heating to the fastest rate possible without elevating the temperature beyond what is needed to kill insects. The potential for movement is restricted by lack of time to reach equilibrium. Slowed cooling by insulating the bag for the cooling phase will not help.

During moisture adsorption a small amount of heat given off by the newly sorbed water moderates the rate of uptake. Full thermodynamic treatment of changes during heating and cooling is complicated by this simultaneous heat and moisture transfer problem. The above description is intended to give the general outline of events, which are also moderated to the point of insignificance for the welfare of the object.

Removal of the bag at the end of active heating, before cooling, results in direct loss of moisture to the atmosphere from both the bag air and the warm object. This would be bad for thin objects, but thicker and painted or gilded objects would 'dry' slowly compared to heat loss and might not suffer unduly.

Leaving the bag on will increase the condensation damage risk, but if the object in the bag is protected or supported, direct contact is avoided and no solution effects would occur. Cotton over-wraps, which cool early and quickly, also absorb vapour and prevent condensation, as well as buffering thin components of objects earlier during heating (Strang, 1999a).

Summary: Time to equilibrium has a principle role to play in heat disinfestation. Slow equilibration favours stability during brief changes in surrounding conditions. Fast equilibration requires some mitigation to prevent unwanted dimensional change, most of which is provided by moisture retention in a polyethylene bag, where heat equilibrates several orders of magnitude faster than moisture. Internal moisture transfer is also limited to comparatively slow mechanisms. Rapid responding dunnage buffers any thin members from moisture loss during the early stages of heat treatment. The condensation risk is strong on cooling, but mitigated by object heat, moisture scavenging cotton over-wraps, and careful object support.

Heat ageing and the risk of damage

Many chemical reactions proceed more rapidly with elevated temperature. Studies on paper, film and magnetic tape binder indicate that the rate of deterioration doubles with increments in temperature of 5°C during the first and most rapid stage of its deterioration (Michalski, 2000). This means elevated temperatures will shorten useful 'lifetime'.

Figure 5 shows the extent of this effect. The inset graph shows the net lowering of expected lifetime of paper objects to 98.6% by an eight-hour heat treatment carried out in perpetuity, every ten years. This is equivalent to the exposure which disinfests entire frame-construction buildings or library stacks. If we expect paper to last at least a century, a single eight-hour exposure will incur something on the order of 0.1% loss of expected lifetime. An hour or so exposure needed for unrolled carpets in the solar bagging application described below, will incur even less deterioration. This cost can be borne when the alternative is destruction by pests.

Figure 6 illustrates how heat affects paper strength. While enduring sharp folding under tension equates to extreme treatment of an object, the comparative data is useful to illustrate how little the paper strength is affected at 60°C when thermal ageing is of short duration. Heat treatment of books would approach eight hours only if one were attempting to heat an entire shelving stack as a unit.

The deterioration of leather increases risk of shrinkage even at temperatures below room conditions when moisture is applied. Irreversible dimensional change in collagen molecules above a threshold temperature (shrinkage temperature, T_s) seriously distorts skin objects. The thermostability of collagen in water relates to the state of deterioration of the object (Young, 1990). Figure 7 shows the temperature for maximum denaturation of collagen against RH (Kopp *et al.*, 1989; Komanowsky,

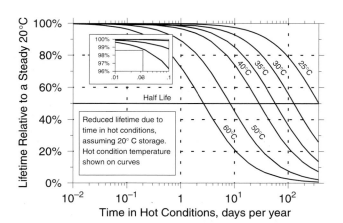

Figure 5 Time in hot conditions greater than 20°C. Data is from Michalski (2000)

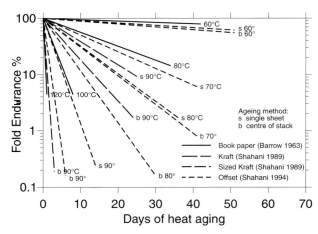

Figure 6 Paper fold endurance after heat ageing. Data is from Shahani (1994), Shahani *et al.* (1989) and Barrow (1963)

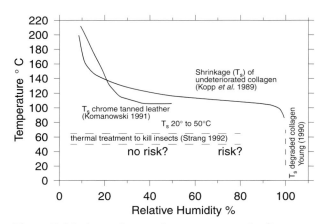

Figure 7 Maximum denaturation temperature of collagen at different relative humidities (Strang, 1995). Data is from Komanowsky (1991), Young (1990) and Kopp *et al.* (1989).

1991). The curve indicates that fresh and tanned collagen at less than 90% RH tolerates temperatures well above 60°C without distortion.

However, problems could arise with deteriorated collagen in ethnographic objects since their shrinkage temperature at saturation can be much lower. Until equivalent curves for deteriorated materials are characterized it is not advisable to subject collagenous material to heat treatment unless its T_s at saturation is above the treatment temperature (Strang, 1995). This point can be determined by microscopic analysis (Young, 1990).

The author performed limited differential scanning calorimetry (DSC) work in determining the T_s of deteriorated collagen under conditions less than saturation, and the results are encouraging. However, the denaturation temperature in the region of 50–70% was above the treatment protocol, and so full characterization is still needed to ensure low overall risk. Thompson's investigation of heat treatment of leather with T_s near 50°C arrived at a similar conclusion (Thompson, 1995). If the material is already damp this might be a case for mild desiccation prior to heat treatment to ensure stability.

At a physical level, melting and softening of waxes, resins and adhesives are a commonly expressed concern when elevating the temperature of objects. Heat above 60°C is commonly used in restoration treatments in a controlled manner to consolidate cupped paintings. Commercial firms use heat to kill insects in buildings and objects as inherently sensitive as Boulle furniture. They do not report damage to shellac and waxed surfaces (Anon., 1990).

Of course, we would not treat objects with materials having melting ranges near 60°C. Resin varnishes and glue become rubbery by 60°C (Michalski, 1991), so caution is advised, but generally this should not pose problems for glued assemblies as the necessary mechanical strain to disrupt joints is not applied in bagged and heated objects.

Summary: The chemical effects of ageing are small enough to allow a discussion of the expenditure of object lifetime for heat disinfestation, as much as we spend on colour by choosing to exhibit under lights, and on lifetime by storing in ambient temperature rather than in cooled rooms.

Water activity, mould risk and condensate distribution

Since moisture movement and its concentration, compounded by the perceived risks from vapour barrier bags is a common concern to those contemplating thermal pest control, some way to estimate subsequent risk by mould is needed. This is extended to storage scenarios simply because bags have a continued use as quarantines before and after heat treatment of collections, certainly until the space is free of pests, and beyond. The microbiological equivalent to relative humidity is the value called water activity (a_w). This term was coined in the food preservation literature 50 years ago to correlate observations on microbial growth to the contribution by water in systems under observation. Water activity is correlated to the ability of water to engage in chemical and biological processes. While study of water activity and microbial processes is complicated by variations and detail of interactions (van den Berg and Bruin, 1981), overall limits are well enough established to provide a useful guide.

The non-ideal (real life) behaviour of water vapour at ambient temperature and pressures is termed fugacity, and is at most 0.2% different from ideal (van den Berg and Bruin, 1981). With allowances for this small difference, water activity is essentially equal to the ratio of measured partial pressure of water to its saturation vapour pressure at a given temperature, and is thus identical to RH expressed as a fraction. Microbiological risk can then be mapped to equilibrium with a system's atmosphere.

The dependence of mould on water activity is summarized in Figure 8. The curve predicts potential for mould growth when humidity (water activity) is known. Higher humidity equates to shorter periods of time before recognizable growth, hence more acute risk occurs. The data in Figure 8 is from one important study on three substrates: dried grass, linseed cake and bone meal,

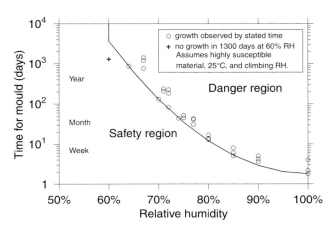

Figure 8 Time for mould growth against relative humidity (Michalski, 1993a). Data is from Snow *et al.* (1944)

representing the most susceptible ethnographic or contaminated objects. The relation between microbial growth and moisture content is a consequence of water activity. We cannot equate growth on one material with potential for growth on another through equivalence in moisture content. We must always refer to the sorption isotherms in order to project risk to other materials, otherwise the prediction is falsely alarmist for mould risk at humidities below 65%.

It is mould and bacterial growth that translates to damage through digestion, or staining. A short time at elevated humidity restricts this potential for damage, so even when heat treatment creates conditions for higher humidity in the bag, mould is not an issue. This is due to the relatively short duration of the resorption phase for mould growth initiation, and the distribution of any condensate on the bag film is well away from the warmer surfaces of the enclosed object (Figure 9). In fact, breaking spore dormancy, only to quickly return to adverse humidity, is beneficial.

The effect of temperature on the need of a mould for moisture is shown in Figure 10, with short curves for old parchment, starched cotton and goatskin in order of increasing sensitivity to mould at lower humidity. The outer boundary is summarized from 50 years of industrial microbiological effort, where moisture content is the key

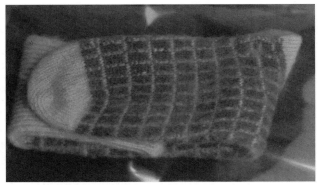

Figure 9 Wool socks at 60°C. Condensate in the bag stays away from the object during cooling

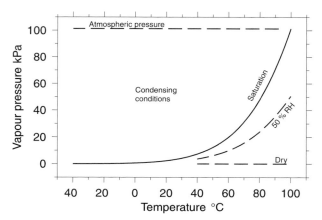

Figure 10 The effect of temperature and humidity for the growth of a mould (Michalski, 1993a)

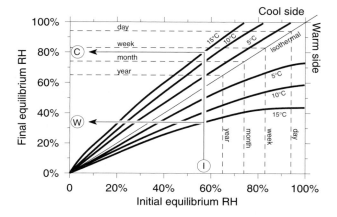

Figure 11 Calculated psychrometric behaviour of bagged wool in a thermal gradient, and mould risk (Strang, 1992). Data is from Hazeu and Hueck (1966) and Michalski (1993a).

for the preservation of goods. Bacteria require greater water activity than moulds. Both high and low temperatures put a strain on organisms in the form of greater dependence on high humidity in order to sustain life. The lesson from this curve is that common moulds, capable of destroying objects, are going to need very high humidities indeed to sustain growth during heat treatment.

Figure 11 shows the effect of equilibrium thermal gradients on moisture distribution in bagged wool (Hazeu and Hueck, 1966) and is similar to cotton. This effect was studied to understand mould risk during ship hold transport of polyethylene bagged materials. Bagged textile, initially at equilibrium with 58% RH (marked I) subjected to a 15°C differential, will drop to 35% RH on the warm side, and climb to 80% RH on the cooler side with a mould risk of a couple of weeks (Strang, 1995).

There are four ways to make an object mouldy in a bag:

- bag at equilibrium RH (ERH) above 65% and maintain long enough for mould to grow (Figure 8)
- bag at an ERH less than 65% but subsequently store at elevated temperature inducing an ERH above 65% (Figure 2)

- bag at ERH less than 65% but store with a thermal gradient significant enough to induce RH above 65% in part of the bag (Figure 11)
- bag at ERH less than 65% but store in air at a higher RH that eventually permeates the bag (Figure 4)

Avoid these situations and the risk of mould is minimal (Strang, 1995).

Summary: The potential for mould growth during thermal pest eradication can be predicted by the RH (equivalent to water activity) of the atmosphere relative to the enclosed materials, irrespective of the materials' moisture contents. Given the short duration of heat treatment, damage by mould is unlikely. Precautions against mould are stated for long-term storage in the bags. Temperature gradients inside the bag must be avoided during long-term storage due to increased mould risk. Examples of risky locations are deep shelves against exterior walls, and bagged or tarpaulined objects resting on concrete floors at grade.

Efficacy of heat disinfestation

In pest control, efficacy is a euphemism for 'ability to kill'. Since the 1900s, efficacy has been a primary requirement for licensing any pesticide, and more recently that it does so without significant side effects. The author performed a literature review of thermal mortality limits to insect populations to establish a solid demonstration of efficacy (Strang, 1992). Fields (1992) published a similar study focusing on agricultural pest control. The summary of thermal mortality data is given in Figure 12 covering 46 species in total, 26 of which are represented in the heat data. The lines show recommended treatment schedules of reasonable duration to ensure efficacy.

The author reviewed the effect of heat treatment on seed viability (Strang, 1999c). Seeds were chosen as a model system, since germination after treatment indicates little chemical or physical effect on a complex living system, unlike the simple and nearly monotonic systems most conservators worry about. Temperature, humidity,

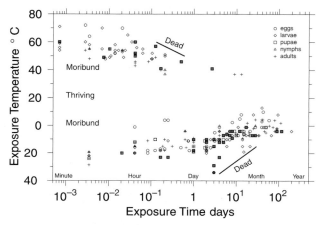

Figure 12 A plot of temperature against time for the mortality of insects (Strang, 1992)

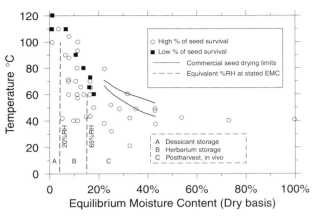

Figure 13 Mortality of seeds correlated to storage temperature, RH and EMC (Strang, 1999c)

and moisture content for optimum seed storage is a well-studied area in agriculture and their recommendations have extensive foundation in annual practice.

By using EMC/RH conversion through the sorption isotherms, seed viability against EMC can be related to viability against RH in storage. Combining viability with temperature and EMC/RH, Figure 13 shows that seeds can undergo heat treatment at 60°C without significant effect on germination up to 60% RH. The effect of low moisture content in reducing damage during heat treatment is clearly seen. For full treatment of thermal control in the herbarium context, see Strang (1999c).

Applications of thermal disinfestation
In the end it all comes down to application. If a technique is too damaging, complicated, expensive or awkward to

employ it will rarely be of value. The principles behind thermal treatments in an efficacious and safe manner are a foundation for developing an application. Application also balances cost, timeliness and availability. All the methods described below have been applied to relieving cultural property of their infestations. The problem lies in the details of implementation.

Demonstration
The simplest demonstration of heating an anisotropic material is to construct a longitudinal grain strip of wood veneer, fused to a cross-grain strip of the same veneer by double-sided adhesive tape. Placed in a polyethylene bag and heated to 60°C with an oven or hair dryer, the strip exhibits very minor curl towards the long grain side, indicative of greater cross-grain thermal expansion. Place the strip in a convection oven, and heat it to 60°C and the strip quickly exhibits extensive curl to the cross-grain side, indicative of massive moisture content loss in both sides and anisotropic response. Figure 14 shows the effect of a 2 L bag around 5 g of wood prepared in the described manner, heated to 60°C for 20 minutes. This is a graphic example of the forces we control with a vapour barrier during thermal treatment; the same forces we know to be responsible for damage to our collections in uncontrolled extreme storage environments.

Solar heating
The author encountered several clients in his advisory role at the CCI who had few resources with which to disinfest

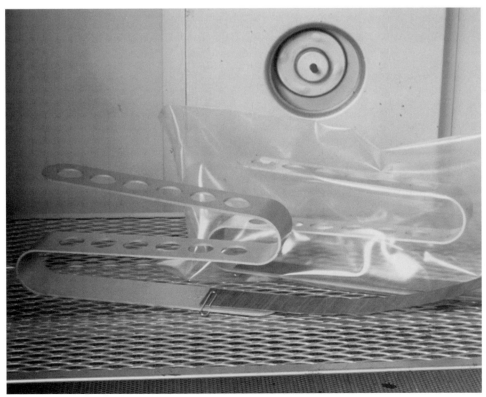

Figure 14 Demonstration of the bag effect during heating to 60°C

textile collections. Clark (1928) demonstrated that sunning textiles killed textile pests. However, exposing dyed fabrics to high light levels and other threats is less acceptable in the museum context. The following system was developed by the author for this more discerning audience, and has been extended by others to wooden objects of the order of 5 cm or less in thickness (Strang, 1995; Strang, 1999a).

Bag a relatively thin object wrapped in cotton in black polyethylene, or place a thin black, well-washed cloth under a clear polyethylene film around the object. A rolled and clipped closure is sufficient if heat sealing is not possible. The bagged object is then placed in a clear plastic surrounding 'greenhouse' to retain heated air outside the black bag and reduce wind cooling. The three following strategies reduce the temperature differential through the objects being treated and the risk of moisture rise on the cooler shade side (as illustrated in Figure 11):

- warm air from the sunny side of the greenhouse is moved to the shady side by mechanical pumping or blowing
- the shady side of the black bag is thermally insulated
- a tubular black skirt is used to absorb heat and conduct it up the shade side

Heating from both sides is superior to an insulated shade side as it speeds treatment and more quickly reduces the thermal gradient.

A black object can exhibit a skin temperature rise of 40°C (Yamasaki and Blaga, 1976), which, on a sunny day in temperate Ottawa, brings the bag to a rapidly lethal 60°C from May to October. This coincides nicely with the opening and closing of seasonal museums which often have resource problems. The black bag protects against light fading and contaminants, as well as the issues detailed in

the above sections. A cotton overwrap between the black polyethylene and the object rapidly delivers initial humidity to the bagged air on heating to protect fine structures, and scavenges moisture quickly on cooling to reduce condensation risks. The equivalent of ten pillowcases per m³ of air should suffice.

This process, detailed in Strang (1999a) was used extensively in field trials between November 1999 and March 2000 by Baskin at the Luang Prabang Museum, in Laos (Figure 15). Heavily infested and damaged shell and glass inlayed lacquered and gilded Buddha figures, as well as fur and fabric items were treated. A solar powered fan was used to blow warm air round to the backside of the black bag. No damage from treatment, condensation or dampness in the cotton wrappings was seen on opening 24–39 hours after the treatment. No sign of infestation was visible after a year (B. Baskin, pers. comm.).

Brokerhof carried out solar heat disinfestations of gilded iconostasis elements in a variety of heat-gathering configurations (Brokerhof, 1998). An L-shaped clear-fronted greenhouse lined with black and insulated from the ground was chosen as the more efficient of the systems tried. The components had been tightly wrapped in black polyethylene, and bag surface temperatures had commonly approached 80°C. Normalized to initial EMC, changes after treatment ranged from +3% to −5% and narrowed over two days from +1% to −3% with an average new EMC, which was 1% lower than initial EMC. Despite the large EMC change in some of the examples, no dimensional change was measured across existing cracks. This indicated that the EMC shift, if as large as was measured, was very shallow as predicted above. Mechanical damage to very tenuously bound surface fragments was noted as equivalent to other handling procedures. No

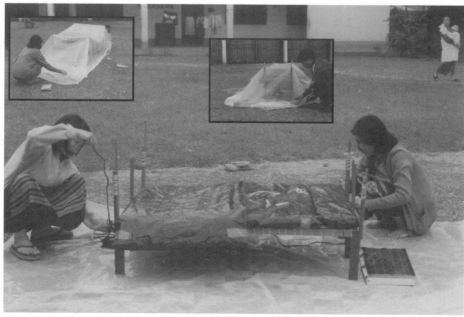

Figure 15 Solar disinfestation system being used to disinfest textiles at the Luang Prabang Museum. Photograph by Baskin (1999)

buffering overwraps had been employed to protect these fine elements from rapid early-treatment EMC shift, or pressure from the bags.

Pearson and his students built a trial solar disinfestation oven, measuring 1 m × 0.5 m² with glass double-glazing and a black lining. Black-bagged objects were set on a rack to allow isothermal heating. Vented with a thermostatically controlled fan operating at 12 V, the chamber maintained a temperature of 58°C to 63°C for three hours, and successfully treated up to 10 kg of black-bagged wood (Strang *et al., 2000*).

Summary: Maximizing solar gain is not the prime objective of this technique, and temperatures high enough to heat the object without soaring past 60°C are sufficient. High skin temperatures prompt surface desiccation and push the bag humidity to unacceptable levels. Ventilation of the outer envelope moderates the inner envelope, and wrapping with cotton moderates surface EMC change and condensation risk. The author's intention for the solar technique was to allow fast and assured pest control to those who have no other choice. However, it need not be considered a last ditch risky method when used with care and understanding of the underlying principles.

Convection heated enclosure

Plant specimens drying at 45°C are at the low end of the lethal boundary. However, an additional situation was proposed by M. Shchepanek of the Canadian Museum of Nature, which was whether it would be possible to heat disinfest material arriving with visiting researchers. The material is generally unmounted, and not likely to go into the permanent collection. Such material does present significant risk of introducing pests, and the social pressures are often one of quick response to the request when a professional colleague appears plant in hand.

The specimens are commonly pressed in folded newspaper, so they can be safely stacked to a thickness of a centimetre between modified herbarium press cardboard. The cardboard acts as both hypocaust and radiator, quickly delivering heat to the centre portion of the specimens (Figure 16).

Heated in a convection oven at 60°C for one hour is sufficient to eliminate pests. No other technique could respond as quickly, with as little risk of insect survival (Strang and Shchepanek, 1995). The specimens were unbagged as additional drying was not a concern. The role of thermal pest control in herbaria has been extensively reviewed by the author (Figure 13; Strang 1999c), as well as its integration into wider IPM strategies (Strang, 1999b).

An insulated plywood box with heat supplied by an internal heater or an external supply can disinfest large or many objects in one treatment. This method was applied to an entire farm equipment collection that was being devastated by woodboring beetles at the Albert County

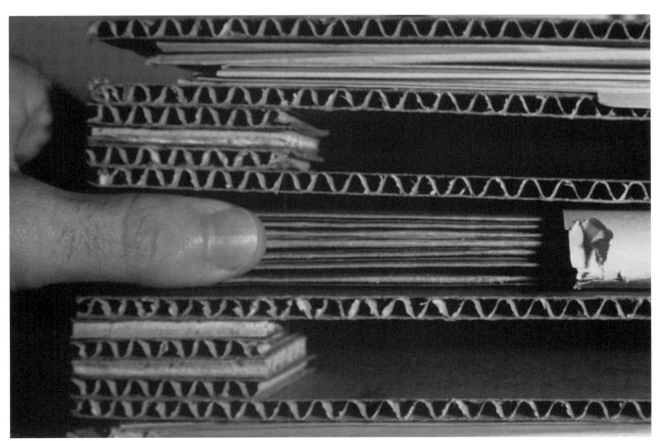

Figure 16 Convection heating herbarium specimens to kill pests, rule-of-thumb for ventilation gap and specimen stack height (Strang and Shchepanek, 1995)

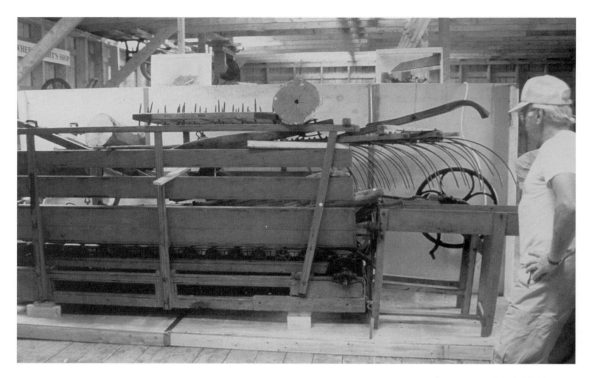

Figure 17 Heat chamber disinfestation of farm machinery. Photograph by Fox (1993)

Historical Society in New Brunswick. A 14 × 4 × 6 foot plywood box with glass-fibre insulation was heated to over 60°C with two 2500 W and 240 V electric heaters, under thermostatic control (Figure 17). A squirrel cage blower ensured air mixing, especially under the raised objects. Temperature measurement within carefully sized blocks of hardwood gave an indication of treatment completion when matched to the largest dimensions of the objects in the chamber.

The system's total expenditure was approximately Canadian $300, carried out by Alastair Fox and society members with initial guidance from the author. Several loads were treated for seven hours each and the infestation eliminated (Fox, 1993). The treatment relied on chamber walls, and the large buffering capacity of the timber objects in recirculated air to keep EMC loss minimal. Bagging, or bagging with cotton over-wraps, is advisable for thinner objects mixed in with larger timbers given the longer treatment time needed for the thicker pieces.

This method, with analysis of effects and efficacy, was carried out by A. Xavier-Rowe and colleagues at English Heritage on treated timber objects associated with

Broadsworth Hall (Xavier-Rowe *et al.*, 2000). In addition, a commercial humidity controlled heat disinfestation chamber is used commercially for heating objects in Europe (Nicholson and Von Rottberg 1996).

Convection heated building

The final and most daunting treatment scale is bagging a building or large outdoor sculpture. The exterior bag reduces reliance on wall insulation and vapour barrier properties, speeds heating and blocks access to the target being heated. Brought to technical pinnacle in the early 1900s for disinfesting granary and milling operations, heat has been employed at building scale throughout the last century. These include:

- heating by hot water radiators scaled to the task (Dean, 1913)
- portable coal-fired furnaces with telescopic pipes to deliver heat to upper stories (Hartnak, 1943)
- portable kerosene fired space heaters (Hartnak, 1943)
- propane-fired heaters with flexible insulating ducts (Forbes and Ebeling, 1987; Nicholson and Von Rottberg, 1996)

These have all been used to eliminate pests from structures and their contents. The treatment time is roughly a working day (can be estimated from Figure 4), with additional time to prepare the site by selective content removal or thermal protection. Tarping the building to allow heating from the exterior wall, thermally insulating interior ground surfaces to avoid cooling and moisture effects, and reversal of all preparations should be taken into account.

While the method can be prone to under-heating portions of the building in contact with the earth, careful setup, monitoring and heat delivery reduces this risk. When woodboring infestation is present in upper reaches, the technique can be very effective. Heat has been applied to fractions of buildings, such as a library within a civic building (Cressman, 1935), leaving other areas largely unaffected by the process. Buildings undergoing heat treatment can be entered for brief periods (Ebeling, pers. comm.) but it is not advisable, as heat stress is a certain risk if any delay is encountered.

CONCLUSION

Temperatures greater than 60°C are used to flatten cupped paintings, set adhesives on textiles, or bake sticky binders onto magnetic tape for one last play. Reasons for using heat therefore run from aesthetic correction, through structural reintegration, ending in salvage from oblivion. The use of heat to control pests in cultural objects is in no way less ethical than these treatments mentioned above.

Pests continually graze the aesthetic surface of our cultural patrimony, leaving ugly scars. Pests undermine the structure, and if uncontested, will reduce the value of the object. Imagine collections exposed in a forest,

and then consider the beneficial effect of a roof and walls. Artificial ageing could equally mean the prolongation of an object's lifetime in the face of the ravages of unchecked nature.

As far as the author knows, the first well-recorded use of heat to disinfest objects for collections is that reported by Kuckahn in 1771 (Kuckahn, 1771). In these communications, Kuckahn details a procedure that ensures the correct temperature for disinfestation of bird mounts, without risk of damage to the feathers. He tested sacrificial feathers in a cooling wood-fired oven until no damage was perceived and then he placed the specimen inside.

> *Baking is not only useful in fresh preservations, but will also be of very great service to old ones, destroying the eggs of insects; and it should be a constant practice once in two or three years, to bake them over again, and to have the cases fresh washed, as above, which would not only preserve collections from decay much longer, but also keep them sweet.*
> (Kuckahn 1771)

Since that time, heat has been applied to seeds prior to planting for fungal control, to kill insects in carpets and furniture, and it has also been used to disinfest houses, granaries, mill structures and libraries. Heat has also protected cultural property, especially in institutions with a very limited budget, lacking electricity, or much pressed by time and volume.

Thermal disinfestation treatment has been introduced to the world many times and it has experienced periods of attention in written records. Ultimately, disinfestation of objects by heat or any other pest control treatment becomes an act of intervention to be gauged relative to doing nothing and accepting the consequences. I would therefore encourage those whose job it is to preserve cultural property to understand the niche and principles of heat disinfestation.

REFERENCES

Andrews G, Dawson R L, 'Ethylene polymers', in *Encyclopaedia of Polymer Science and Engineering*, Kroschwitz J I (Editor), John Wiley and Sons, 1986, volume 6, 383–522.

Anon., 'Konservierungstechnik – thermische vernichtung holzzerstorender insecten', *Restauro*, 1990, **96**(2), 80.

Anon., Mackintosh J G (Editors), 'Psychrometrics', in *ASHRAE Handbook, Fundamentals*, SI edition, chapter 6, 1985, American Society of Heating, Refrigerating and Air-Conditioning Engineers, Atlanta, USA.

Anon., Parsons R A (Editors), 'Museums, libraries, and archives', in *ASHRAE Handbook, Heating, Ventilating, and Air-Conditioning Applications*, SI edition, chapter 20, 1999, American Society of Heating, Refrigerating and Air-Conditioning Engineers, Atlanta, USA.

Barrow W J, *Permanence/Durability of the Book. A Two Year Research Program*, 1963, W J Barrow Research Laboratories, Richmond, Virginia, USA.

Brokerhof A W, 'Pest control treatment of the Probota iconostasis, part 2: heat treatment of the decorative parts', 1998, *UNESCO Project 536/ROM/70.*

Bry R E, McDonald L L, Lang J H, 'Protecting stored woollens against fabric-insect damage: a long term nonchemical method', in *Journal of Economic Entomology,* 1972, **65**(6), 1735–1736.

Clark C O, 'The protection of animal fibres against clothes moths and dermestid beetles', in *Journal of the Textile Institute,* 1928, **19**(3), 295–320.

Crank J, 'Rate of change of moisture content', in *Moisture in Textiles,* Hearle J W S and Peters R H (Editors), 1960, Butterworths, New York, 83–93.

Crank J, *The Mathematics of Diffusion,* 2nd edition, 1975, Oxford University Press, London.

Cressman A W, 'Control of an infestation of the cigarette beetle in a library by the use of heat', in *Journal of Economic Entomology,* 1935, **26**, 294–295.

Dean G A, 'Methods of controlling mill and stored-grain insects, together with the habits and life histories of the common infesting species', in *Mill and Stored-Grain Insects,* 1913, no. 189, Kansas State Agricultural College, Agricultural Experiment Station, Manhattan, KS, USA, 139–236.

Fields P G, 'The control of stored-product insects and mites with extreme temperatures', in *Journal of Stored Product Research*, 1992, **28**(2), 89–118.

Forbes C, Ebeling W, 'Update: use of heat to eliminate structural pests', in *The IPM Practitioner,* 1987, **9**(8), 1–5.

Forest Products Division Forestry Branch, *Canadian Woods, Their Properties and Uses,* 2nd edition, Baxter A J, Potter G R L, Schryburt (editors), 1951, King's Printer, Ottowa, Canada.

Fox A, *A method of using heat to treat wood-boring insects at the Albert County Historical Society,* 1993, Report to the Canadian Conservation Institute, Ottawa, Ontario, Canada.

Gál S, 'Recent advances in techniques for the determination of sorption isotherms', in *Water Relations of Foods,* Duckworth R B (Editor), 1975, Academic Press, London, 139–154.

Harriman L G (Editor) *The Dehumidification Handbook,* 2nd edition, 1990, Munters Cargocaire, Amesbury, MA, USA.

Hartnak H, *Unbidden House Guests,* 1943, Hartnack Publishing Company, Tacoma, Washington DC, USA.

Hazeu W, Hueck H J, 'Changes of humidity inside packages due to environmental conditions', in *Microbiological Deterioration in the Tropics,* 1966, Society of Chemical Industry, London.

Highland H, 'Insect infestation of packages', in *Insect Management for Food Storage and Processing,* Baur F J (Editor), 1984, American Association of Cereal Chemists, Minnesota, USA.

Komanowsky M, 'Thermal stability of hide and leather at different moisture contents', in *Journal of the American Leather Chemists' Association,* 1991, **86**, 269–280.

Kopp J, Bonnet M, Renou J P, 'Effect of collagen crosslinking on collagen–water interactions (a DSC investigation)', in *Matrix,* 1989, **9**, 443–450.

Kuckahn T S, 'Four letters from Mr. T. S. Kuckhan [sic], to the president and members of the Royal Society, on the preservation of dead birds', in *Philosophical Transactions,* 1771, **60**, 302–320.

Latta J K, Beach R K, 'Vapour diffusion and condensation' in *Canadian Building Digest,* 1964, **57**.

Michalski S, *Correlation of zero-span strength, fold endurance, RH and temperature in the ageing of paper: a review of published data,* 1990, Internal report, Canadian Conservation Institute, Ottawa, Ontario, Canada.

Michalski S, 'Paintings – their response to temperature, relative humidity, shock, and vibration', in *Art in Transit, Studies in the Transport of Paintings,* Mecklenburg M F (Editor), 1991, 223–248.

Michalski S, 'Relative humidity: a discussion of correct/incorrect values', in *Pre-prints of ICOM Conservation Committee 10th Triennial Meeting,* 1993a, 624–629.

Michalski S, 'Relative humidity in museums, galleries, and archives. Specification and control', in *Bugs, Mold and Rot II,* 1993b, National Institute of Building Sciences, Washington, USA, 51–62.

Michalski S, 'Leakage prediction for buildings, cases, bags and bottles', in *Studies in Conservation,* 1994, **39**, 169–186.

Michalski S, 'Quantified risk reduction in the humidity dilemma', in *APT Bulletin,* 1996, **27**(3), 25–29.

Michalski S, *Relative Humidity and Temperature Guidelines for Canadian Archives,* 2000, Canadian Council for Archives, Ottawa, Canada.

Mueller M F, 'Psychrometric behaviour in closed packages', in *Modern Packaging,* 1949, **22**(11), 163–167.

Newton J, 'Insects and packaging: a review', in *International Biodeterioration,* 1998, **24**, 175–187.

Nicholson M, Von Rottberg W, 'Controlled environment heat treatment as a safe and efficient method of pest control', in *Proceedings of the 2nd International Conference on Insect Pests and Urban Entomology,* Edinburgh, 1996, 263–265.

Pauly S, 'Permeability and diffusion data', in *Polymer Handbook,* volume VI, 3rd edition, Brandrup J and Immergut E (Editors), 1989, Wiley-Interscience, 435–449.

Richard M, 'Control of temperature and relative humidity in packing cases', in *Art in Transit, Studies in the Transport of Paintings,* Mecklenberg M F (Editor), 1991, 279–297.

Shahani C J, 'Accelerated aging of paper: Can it really foretell the permanence of paper', in *Proceedings of the Workshop on the Effects of Aging on Printing and Writing Papers,* July 1994, PCN: 33-000009-11, ASTM Institute for Standards Research, Philadelphia PA, USA, 120–139.

Shahani C J, Hengemihle F H, Weberg N, 'The effect of variations in relative humidity on the accelerated aging of paper', in *Historic Textile and Paper Materials II*, Zeronian S H and Needles H L (Editors), 1989, American Chemical Society, Washington DC, USA.

Snow D, Crichton M H G, Wright N C, 'Mould deterioration of feeding stuffs in relation to humidity of storage', in *Annals of Applied Biology*, 1944, **31**, 102–110.

Strang T J K, 'A review of published temperature for the control of pest insects in museums', in *Collection Forum*, 1992, **8**(2), 41–67.

Strang T J K, 'The effect of thermal methods of pest control on museum collections', in *Biodeterioration of Cultural Propery*, 1995, **3**, 334–353.

Strang T J K, 'Solar heating to eliminate insect pests in collections', Draft Note, 1999a, Canadian Conservation Institute, Ottawa, Ontario, Canada.

Strang T J K, 'A healthy dose of the past: A future direction in herbarium pest control?', in *Managing the Modern Herbarium,* 1999b, 59–80.

Strang T J K, 'Sensitivity of seeds in herbarium collections to storage conditions, and implications for thermal insect pest control methods', in *Managing the Modern Herbarium*, 1999c, 81–102.

Strang T J K, Shchepanek M, *Fast elevated temperature control of insects in vascular plant specimens,* Poster at the Society for Preservation of Natural History Collections, Annual Meeting, 2–4 June 1995, Toronto, Canada.

Strang T J K, Mitchel J, Pearce A, Pearson C, *Low cost methods for insect pest control,* 2000, Poster at the International Institute for Conservation, Congress, Melbourne, Australia.

Thompson G, 'Relative humidity-variation with temperature in a case containing wood', in *Studies in Conservation,* 1964, **9**, 153–169.

Thompson R S, The effect of the Thermo Lignum® pest eradication treatment on leather and other skin products, in *ICOM Leathercraft Group Conference,* Hallebeek P B and Mosk J A (Editors), April 1995, ICOM Committee for Conservation, Netherlands Institute for Cultural Heritage, Amsterdam, 67–76.

Toner R K, Bowen C F, Whitwell J C, 'Equilibrium moisture relations for textile fibers', in *Textile Research Journal,* 1947, **17**, 7–18.

Urquhart A R, 'Sorption isotherms', in *Moisture in Textiles,*

Hearle J W S and Peters R H (Editors), 1960, Butterworths Scientific Publications, London, 14–32.

van den Berg C, Bruin S, 'Water activity and its estimation in food systems: theoretical aspects', in *Water Activity: Influences on Food Quality*, Rockland L B and Stewart G F (Editors), 1981, Academic Press, London.

USDA, *Wood Handbook,* revised edition, 1987, US Department of Agriculture, Washington DC, USA.

Weast R C (Editor) *Handbook of Chemistry and Physics,* 53rd edition, 1972–3, Chemical Rubber Company, Cleveland, Ohio, USA.

Xavier-Rowe A, Imison D, Knight B, Pinniger D, 'Using heat to kill museum insect pests – is it practical and safe?', in *Tradition and Innovation, Advances in Conservation,* International Institute for Conservation (IIC), 10–14 October 2000, 206–211.

Yamasaki R S, Blaga A, 'Hourly and monthly variations in surface temperature of opaque PVC during exposure under clear skies', in *Materiaux et Constructions,* 1976, **9**(52), 231–242.

Young G S, 'Microscopical hydrothermal stability measurements of skin and semitanned leather', in *Pre-prints of the ICOM Committee for Conservation, Los Angeles,* 1990.

ACKNOWLEDGEMENTS

The author would like to thank Stefan Michalski, Charlie Costain, Robert Waller, Greg Young, Scott Williams, Roy Thompson, Monika Harter, Mike Shchepanek, Walter Ebeling, Paul Fields, Bonnie Baskin, Agnes Brokerhof, Alastair Fox, Colin Pearson, David Pinniger, and Amber Xavier-Rowe for their contributions and communications with the author on heat disinfestation treatment over the years.

BIOGRAPHY

Tom Strang is a research scientist with the Canadian Conservation Institute (CCI) in Ottawa, Canada. For the last decade he has investigated the efficacy and utility of thermal and controlled atmospheres to assist in their acceptance into the conservation community as routine methods of pest control to replace proscribed fumigants. He has also participated in conservation aspects of shipwrecks, historic site development, museum exhibits, natural history collections, and lead GPS survey and GIS work on high arctic fossil forests.

CHAPTER NINETEEN

BATTLE OF THE BEASTS: TREATMENT OF A PEST INFESTATION OF THE MOUNTED MAMMAL COLLECTION AT LIVERPOOL MUSEUM

Janet Berry

Conservation Centre, National Museums and Galleries on Merseyside, Whitechapel, Liverpool L1 6HZ, United Kingdom★

★Author's current address: Department of Museum Studies, Leicester University, 103–105 Princess Road East, Leicester LE1 7LG, United Kingdom Tel: +44 116 252 3963

ABSTRACT

The National Museums and Galleries on Merseyside (NMGM) 2001 project has involved the movement of much of Liverpool Museum's collections to temporary storage facilities whilst parts of the museum are refurbished. It was during this movement of collections that an infestation was discovered in one of the mounted mammal collection stores.

The culprit was identified as *Anthrenus sarnicus* (Guernsey carpet beetle). A team of conservators, curators and a taxidermist surveyed the damage and drew up an action plan for a counterattack offensive on the *Anthrenus*. This involved treatment by freezing in the Conservation Centre cold room, which contained 205 mammals ranging in size from a desert rat to a polar bear, as part of a rolling programme.

This paper discusses how the treatment plan was put into action, the practicalities in terms of staff time and resources within the context of the NMGM 2001 project, the rationale behind freezing as a treatment and monitoring after the event.

KEYWORDS

Mounted mammals, *Anthrenus sarnicus*, infestation, freezing treatment

INTRODUCTION

The Liverpool Museum, National Museums and Galleries on Merseyside (NMGM), have a natural history collection of international significance. As part of the Heritage Lottery funded NMGM 2001 project, the museum has recently undergone major redevelopment. Before this, much of the natural history collection was stored in spaces that were originally used for display such as the Horseshoe Galleries. This area of Liverpool Museum had suffered from wartime damage and inadequate post-war reconstruction (NMGM, 1997). The 2001 project involves the restoration of these galleries for public displays, and the creation of new storage areas in the basement of the building for much of the natural history and science collections.

The mounted mammals were one of the collections stored in the Horseshoe Galleries. This area was an open-plan space with a mix of collections and curatorial staff offices. Many of the mounted mammals were stored in a temporarily partitioned room, but the mounted birds in glass display cabinets and the larger specimens, such as the reposing lion and pouncing lioness, were kept in the outer area. Specimens inside the room were stored on open shelves or on their bases directly on the floor. As the collection grew with new acquisitions, the storeroom became more densely packed, making access to specimens very difficult. With the densely-packed store room and open storage of the mammals, the risk of insect infestation was very high, and indeed there had been previous infestations of *Anthrenus*.

With the refurbishment of the galleries, the storage problems for the collection were finally addressed. The smaller mounted mammal specimens were to be housed in a new store in the basement of the main Liverpool Museum building and larger specimens were to be moved to a room in another of NMGM's stores at North Street. All of the collection would be in enclosed stores and separated from other collections, thus reducing the risks of infestation.

However, as both the galleries and the basement areas in the museum were to be refurbished at the same time, it was necessary for collections in the Horseshoe Galleries to be decanted either to North Street or to a temporary store in north Liverpool (Bootle). Therefore, the smaller mounted mammals would be stored with the bird, entomological and botanical collections for a period of up to two years.

As the collections were to be moved out, they had to be physically checked and wrapped for the decant. It was during this checking that a new infestation of *Anthrenus sarnicus* was discovered in the mammal store.

THE BATTLE CRY IS SOUNDED

Parts of the collection had previously suffered from repeated infestations of *Anthrenus sarnicus*, most recently in 1997. After this the room's floor and hessian ceiling had been sprayed with Ficam W® and a programme of monitoring using insect traps had been instigated. However, because of the overcrowded storage of mammal specimens, it had not been possible to check the objects for infestation and a new infestation had taken hold.

RAPID RESPONSE UNIT

When the infestation was discovered, the curators alerted the relevant conservation sections and taxidermy, and a team was established to deal with the problem. The initial response was to:

- assess the extent of the infestation
- assess the damage to objects
- isolate infested material
- arrange treatment
- alert other staff in the building to the presence of the infestation

The collection in the store was examined, and divided into objects with apparent active infestation and those without. The insect pests were identified by the Entomology section at the museum as *Anthrenus sarnicus*. Different stages in the life cycle were found on the objects, including different larval instars, cast skins and adults. This indicated that the infestation was well established (Hillyer and Blyth, 1991). Infestation was concentrated on specimens stored near an internal window between the storeroom and the main gallery, and on small mammals stored on a shelf and within a 0.5 m radius around the bases on the floor (concentrating on feet and trailing tails), i.e. within crawling distance for the larvae. Twenty specimens had evidence of heavy and active infestation.

As the infestation was near a window connected to the main gallery area, the collections in the surrounding areas had to be checked for signs of infestation.

The next step was to organize a programme for treatment of the material. As the work progressed, it became apparent that objects in the mammal store would have to be moved to the temporary store before treatment could be undertaken. As the objects would have to be in the same area as other natural history collections, the risk of spreading the infestation was very high. Therefore, it was decided that all the specimens in the area needed to be treated, to ensure that any eggs or larvae missed in the initial search would be killed. This entailed treatment of 125 mounted mammals and 80 mounted game heads, a total of 205 specimens.

The most severe cases could be treated first, whilst the rest would be wrapped to isolate them from the other collections.

THE SECRET WEAPON

When the Conservation Centre was built, it was designed with a quarantine room containing a walk-in cold room. The cold room is able to treat large objects for pest disinfestation treatments, and reaches temperatures of −29°C in two to four hours. The room is positioned on the ground floor next to the loading bay, to allow potentially infested material to be brought in without risk to other material in the building.

The freezing process has been used as a pest disinfestation technique for centuries (Florian, 1997). The effect of low temperature on insect mortality has been the subject of literature reviews (for example, Strang, 1992; Florian, 1997), in the search for the most effective eradication process for museum insect pests. An important part of the process is to ensure low temperatures are reached quickly, to prevent cold-hardening of insects. Whilst there is little information in the literature that museum insect pests exhibit a significant degree of freeze-tolerance that would prevent death below −20°C (Strang, 1992; Brokerhof and Banks, 1993), recommendations are that measures should be taken to prevent cold-hardening occurring during the freezing process (Florian, 1997).

At the Conservation Centre, the procedure is to put material into the cold room at room temperature, set the target temperature to −29°C and maintain this temperature for 72 hours. This is followed by a slow rise to room temperature over 24–48 hours. Experiments have demonstrated that temperatures at the core of objects reach −29°C in 9–12 hours (T Seddon, personal communication). For a successful freeze, objects have to be sealed in polythene, to prevent moisture loss and condensation during the freezing process (Florian, 1997) and at NMGM, parcel tape is used to seal the polythene.

It was therefore decided to treat the collection by freezing.

Ideally, the objects would have been sealed, transported to the Conservation Centre in batches for freezing, and then sent to their temporary or permanent stores. However, the schedule for the movement of the collections from the museum for the start of building works did not allow for any delay in the decant process.

Our immediate priority then became the isolation of the specimens that had potentially active infestations before they were moved, which involved ensuring the objects were sealed in polythene before moving.

Figure 1 Wrapped game heads are placed in the cold room on open metal shelving in preparation for the freezing treatment

Figure 2 Six people were required to move the pouncing lioness into the new mammal store at North Street after the freezing treatment

The decant schedule did allow for some of the worst-affected mammals to be wrapped, transported and frozen. Therefore, an emergency freezing schedule was arranged for the highest priority material. The rest of the specimens were transported to the temporary stores, awaiting the availability of the Conservation Centre Handling and Transport team.

The Handling and Transport team works to provide in-house transportation of objects. The number of specimens frozen at one time depended not only upon the capacity of the cold room, but also the capacity of the team's van. In general, between 10 and 25 mammals were frozen in one treatment, depending upon the size of specimen. Mixes of small and larger specimens were combined to make optimum use of the racking and floor space in the cold room (Figure 1). Larger specimens were placed on pallets to allow air flow around their base and prevent warm pockets occurring. In total, 12 freezing treatments were undertaken. The associated moves (2 moves for each treatment) had to be scheduled with the Transport and Handling timetable, including extra staff at times for the movement of the larger items (Figure 2).

The cold room was fitted with a ramp, which made movement of larger objects into and out of the room easier (Figure 3). Although the internal dimensions of the cold room are large (3.82 m wide x 2.80 m long x 2.30 m high),

Figure 3 A ramp on the door into the cold room allowed the reposing lion to be wheeled directly into the room from the loading bay by two people

Table 1 Number of hours spent treating infestation in addition to packing of the collection for decant

Procedure	Hours spent[1]
Initial assessment and emergency measures	72
Additional wrapping or sealing of objects for freezing[2]	144
Transport and handling to or from Conservation Centre cold room	134
Rearrangement of specimens in store	27
Associated paperwork by conservation staff	24
Total	**491**

[1] All times are based on an average eight hour day with one hour lunch and two half-hour teabreaks, giving a working time of six hours

[2] Some of the specimens were wrapped for transport by an external contractor. This figure is for additional wrapping required to prepare specimens for freezing

the limiting factor is the door size (1.00 m wide x 1.99 m high). Although animals such as gazelles were themselves quite slender, their bases had to be taken into consideration. For some mammals with bases wider than the door, extra help had to be sought to carefully manoeuvre objects into the room.

RESOURCES

As already stated, the mammals were due to be packed and moved as part of the 2001 project. However, this temporary wrapping was not sealed to the standards needed for freezing, so more time and expense had to be spent on wrapping. Additional time was also required for the initial assessment, freezing treatments and movement of the specimens. Table 1 gives a breakdown of the amount of time spent by staff dealing with the infestation, and Figure 4 presents the times as percentages. The total of 491 person hours represents approximately 2.4 hours per specimen treated.

The main staff involved in the process were two conservators, three curators, one taxidermist, and six members of the Handling and Transport team, plus administration and security staff.

It can be seen from Table 1 and Figure 4 that a large proportion of time was necessary for wrapping the collections prior to treatment. This figure would have been larger if the mammals had not already been wrapped for decant. Transport and handling took the second largest proportion of time. This is high due to a number of the specimens that required six people to move them, making the moves labour-intensive. Therefore, although each move took on average two hours, this involved a total of 12 person hours per move for eight of the moves (four treatments). The percentage of time spent on organization of the stores is under-represented, as the collection in temporary storage has not yet been moved back to the main museum. More time will be spent by curators and conservators ensuring that the collections are organized for

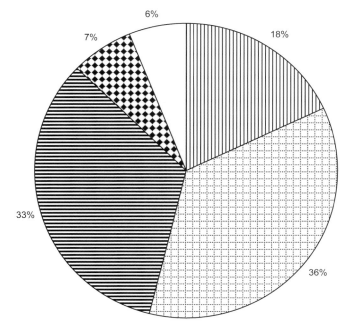

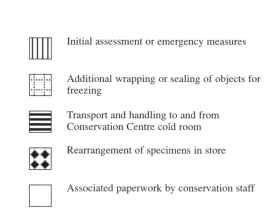

Figure 4 Proportion of time spent by staff on infestation

better protection against reinfestation than under previous conditions.

Each freezing treatment takes approximately five and a half days (12 hours to reach temperature, 72 hours at constant temperature and 48 hours to return to room temperature). For the 12 processes undertaken, this makes a total of 66 days for the entire freezing process. The cost of electricity for the running of the cold room for 72 hours at −30°C is approximately £50 (based on UK business rate electricity tariffs). Therefore, the total electricity cost for the treatment of 205 objects was £600 (excluding labour and packing materials).

AFTERMATH

After the freezing treatment, the objects taken to the temporary store in Bootle are kept wrapped in polythene until they are moved to their permanent store in the newly refurbished museum, with a trap monitoring programme to check for signs of new infestations.

The large mammals and mounted game heads have been placed permanently in a separate store and four curators, one taxidermist and one conservator spent a day unwrapping, removing old evidence of infestation and mounting the game heads on the wall. The larger mammals have been kept wrapped to inhibit reinfestation from any external sources. The room has insect monitoring traps, and the objects are spaced out to allow access for regular checks.

Much of the movement of the collection would have occurred during decant, but the extra time and resources needed for the freezing process were a significant addition to our workload. With the schedules of staff involved, the process took nine months to complete. New, better-designed stores will vastly reduce the chances of reinfestation and with better object access, checking for infestations will be much easier.

NMGM now has an IPM policy document, which clearly defines roles relating to the monitoring and treatment of pests. This paper shows the amount of time and resources needed to treat infestations. This was one battle won in the war against insect pests in museums.

REFERENCES

Brokerhof A W, Banks H J, 'Cold tolerance of *Tineola bisselliella* (Lepidoptera: Tineidae) eggs at a slow cooling rate', in *Journal of Stored Products Research*, 1993, **29**(4), 305–310.

Florian M-L, *Heritage Eaters. Insect and Fungi in Heritage Collections*, 1997, James & James (Science Publishers), London.

Hillyer L, Blyth V, *Carpet Beetle – A Pilot Study in Detection and Control*, 1991, Victoria and Albert Museum Internal Report, London.

NMGM, *NMGM 2001: a celebration of the links between art, history and science. An outline bid to the Heritage Lottery Fund on behalf of the Board of Trustees of the National Museums and Galleries on Merseyside*, 1997, National Museums and Galleries on Merseyside Internal Document.

Strang T J K, 'A review of published temperatures for the control of pest insects in museums', in *Collection Forum*, 1992, **8**(2), 41–67.

MATERIALS AND EQUIPMENT

Ficam® W is a non-repellent odourless water dispersible powder containing 80% w/w 2,2-dimethyl-1,3-benzodioxol-4-yl methylcarbamate (Bendiocarb). It was applied as a spray in water following relevant health and safety procedures.

Killgerm Chemicals Ltd
Denholme Drive
Ossett
West Yorkshire WF5 9NA
United Kingdom
Tel: +44 1924 277631
Fax: +44 1924 264757

ACKNOWLEDGEMENTS

The successful treatment of this infestation could not have been achieved without the enormous help of Tracey Seddon, Clem Fisher, Tony Parker, Malcolm Largen, George McInnes and the Conservation Centre Handling and Transport team. Chris Felton provided identification and interesting intimate details on the life cycles of insect pests which were caught. Thanks go to the team above and to Siobhan Watts and Vivien Chapman for advice with this paper.

BIOGRAPHY

Janet Berry was the Environmental Control Officer for NMGM from May 1999 to August 2001, based in the Conservation Science and Research Section at the Conservation Centre. In January 2000, she took over the role as Pest Management Coordinator from Tracey Seddon. Janet studied archaeological conservation at Cardiff University before undertaking research into stone conservation at the Institute of Archaeology, University College London, where she was awarded an MPhil. She has worked as a conservator on sites and in museums in Europe, the Middle East and USA. Before joining NMGM, she worked for the National Trust specialising in environmental control within historic properties. In September 2001, Janet moved to Leicester University to lecture on Collections Care in the department of Museum Studies.

CHAPTER TWENTY

COLLECTION IN PERIL: INSECT PEST ERADICATION IN ETHNOLOGY STORAGE AT THE ROYAL ONTARIO MUSEUM IN CANADA

Elizabeth C Griffin

5 Wyandot Avenue, Toronto, Ontario M5J 2E6, Canada
Tel/Fax: +1 416 203 1015   e-mail: elizabeth.griffin@primus.ca

ABSTRACT

During 1997–98 an intensive project was undertaken at the Royal Ontario Museum in Toronto, Canada, to eradicate insects in ethnology storage. This paper describes the remedial action taken in response to an escalating number of webbing clothes moth (*Tineola bisselliella*) sightings in the storage areas. The entire collection of ethnographic objects in storage was systematically examined, packed and treated with cycled freezing or fumigation. Alternate methods were developed for culturally sensitive materials. Throughout the project, numerous moth infestations were found in all types of organic materials and woodboring beetles were discovered in large African sculptures. To maintain accessibility to the collection which was frequently used by researchers and curators, and to ensure isolation of treated artefacts, were important criteria. New storage mounts were developed to protect objects during transfer. This paper describes the practical logistics, methodology and observations made during this salvage operation. Post-project preventive strategies are also reviewed.

KEYWORDS

Anthropology, webbing clothes moth, cycled freezing treatment, packing, pest control

BACKGROUND OF THE PROJECT

The Royal Ontario Museum (ROM) ethnology collection comprises North American aboriginal artefacts, including an extensive collection of Arctic and subarctic materials, as well as African, Polynesian, Melanesian and South American artefacts. Objects are housed in a large storage area located within the Anthropology department. The main ethnology store is one of the largest in the museum building, housing 43,000 objects in 474 m² of usable space. Most of the collection is stored on open metal shelving units or in wooden cabinets (Figure 1). Insect sightings, first recorded in 1996, were cause for alarm. The immediate threat to the collection was a webbing clothes moth (*Tineola bisselliella*) infestation of unknown proportions. A project to eradicate insect pests from the storage space and artefacts was initiated in April 1997 and continued until May 1998. The handling of the wide range of diverse objects posed interesting challenges in terms of stabilization and treatment. Objects to be treated ranged in scale from massive West Coast totems, a full-size reed boat from Lake Titicaca, to minute feather ornaments.

PLAN OF THE PROJECT

As the source and extent of the infestation were unknown, the storage room and its contents had to be dealt with in its entirety. The approach taken involved a systematic examination and wherever possible, treatment of all objects to arrest damage to artefacts and prevent the spread to other parts of the museum.

The cleaning of the storage furniture and room infrastructure was also necessary to rid the space of insects. Eggs may be deposited on inorganic materials, including plastic and metal, whilst the destructive larvae congregate on food sources. Flooding the storage room with a chemical gas that would be safe for artefacts and humans was not a feasible option. Other methods, such as intensive trapping with pheromone lures to eliminate flying male insects, would not have been sufficiently effective in terms of quantity and killing all species.

Therefore, a systematic approach was used in order to eliminate the possibility of missing an infestation site and to make the remedial processes as efficient as possible. Strategies were developed and refined with the aim of minimizing the amount of handling required for each artefact.

Due to resource and time restraints, cycled freezing was chosen as the primary form of treatment. For some materials fumigation with phosphine, from a metallic phosphide Phostoxin™, was implemented. The general procedures were as follows:

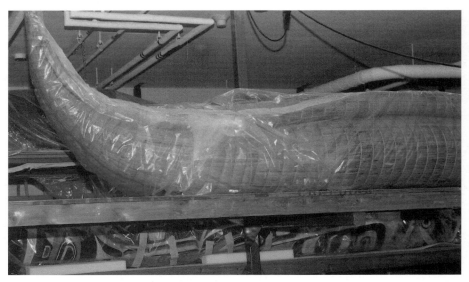

Figure 1 Reed boat on metal shelving in ethnology storage

- artefacts were examined, packed, documented and re-shelved (Figure 2)
- packed objects were transferred in sequence to the freezers on a daily basis or to a fumigation chamber on a bi-monthly schedule
- after treatment was implemented objects were returned to their original place in storage, with the packing still in place

The two important criteria in the plan were that the collection remained accessible to enable the usual research and curatorial work to continue, and to ensure that treated materials were isolated from the contaminated storage room. In addition, the need to stabilize artefacts for transit and treatment became a significant part of the process. The treatment programme required major considerations in terms of coordinating the movement of a huge volume of artefacts between the treatment sites and the ethnology store. Objects were moved either to the freezers, located on the same floor as the storage room, or to the fumigation chamber in the basement of the museum, which was reached via the freight elevator.

IMPLEMENTATION

Preparation

The artefacts were removed from shelves and drawers in the sequence in which they were to be examined and packed at a workstation equipped with a magnifying light. As work progressed, the workstation was moved to different sections of the storage room, which had been subdivided by temporary walls of plastic sheeting in an attempt to contain infestations.

Figure 2 Preparation of objects for treatment

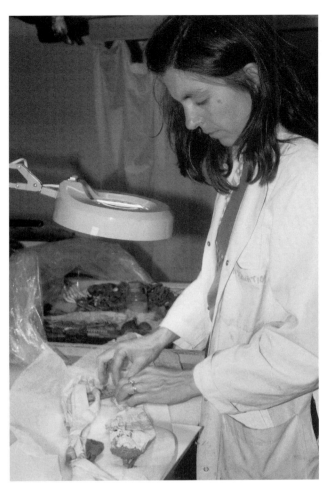

Figure 3 Examination of artefacts

Examination

The initial procedure involved examination for two purposes: to search for evidence of insects, and to assess artefact structure, condition and materials in order to select the appropriate treatment (Figure 3). Storage furniture was examined for signs of insects and cleaned.

In addition to live and dead larvae and adults, clear signs of recent insect activity in artefacts were the findings of larval moults, undisturbed quantities of frass (associated with flight holes in wood) and 'fresh' webbing debris and cocoons. Cocoons were sometimes camouflaged by the integration with the artefact material. The presence of insects was also indicated by evidence of attack, which took various forms including loosened hair or fur under which larvae had tunnelled, matted and missing fibres, staining, structural holes, chewed edges and sometimes visible channelling made by woodborers (Figure 4).

Examination was complicated by the presence of insect debris and damage sustained in previous infestations. However, as examinations progressed, a number of clues indicated the scope of the current infestation. Adult and larval specimens, and some moults and frass were sent to the Entomology department for identification, when the species was unknown to conservation staff.

Packing and stabilization of artefacts for transit and treatment

The quantity of artefacts meant that the packing processes had to be as efficient as possible while ensuring the safety of the object during transit and treatment. Objects as diverse as totem poles and hide shields had to be prepared for transit within the course of a single day. Although the artefacts did not travel very far, various levels of stabilization were needed for many types of objects beyond the normal cushioning. Supports to protect artefacts from physical stress and mechanical damage were essential. The supports and braces, which protected objects in transit, were retained to serve the same function in storage after treatment.

Figure 4 Webbing clothes moth infestation, on a goatskin bag from Pakistan

Design criteria for supports included:

- the provision of stabilization and a safe and obvious means of handling
- ease of fabrication or assembly
- the use of non-reactive materials
- minimum visual interference with object
- dimensions compatible with storage space
- consolidation of numbers of small objects into a more portable form

The supports varied from the simple inclusion of a rigid base padded with Ethafoam®, to the fabrication of external armatures (Figures 5 and 6). Certain types of objects required more extensive preparations due to cracked or weakened structures, fragile components or friable surface coatings. These included African masks, arrows with feathers, headdresses and some large three-dimensional fibre structures such as baskets. Flat objects

such as textiles, matting, bark cloths, feather ornaments and leather garments with stitched beadwork were moved on supports, which was particularly useful when removing objects from the freezers for the thaw cycles. Mounts or padding were also used with artefacts to provide support for weakly attached appendages and heavy decorative elements, in order to restrict movement and to act as a buffer against vibration (a feature of travel in the freight elevator). Delicate three-dimensional structures required simple armatures to protect against the sagging of plastic sheeting (Figure 7). A limited number of materials were used.

Ethafoam® sheeting and acid free board materials were cut to size with utility knives, with off-cuts also providing bases. Ethafoam® blocks placed as strategic supports were anchored with twill tape ties to Coroplast® (museum mounting board) bases. Polyester batting, clear polyester film, steel staples and brass fasteners were occasionally

Figure 5 Support and headdress from the North West Coast, made from wood, ermine, abalone and eagle down

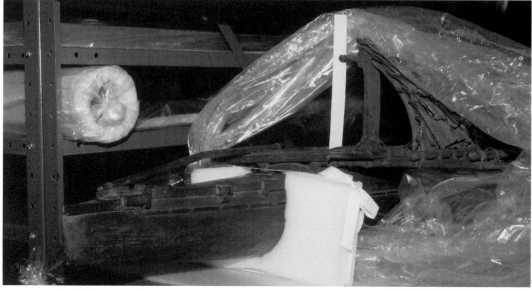

Figure 6 Packing and wooden boat model from Polynesia

Figure 7 Packing and basket from the North West Coast, made from cedar and cherrywood bark

useful. Adhesives were not used. Chemically inert materials were used rather than temporary disposable packing materials to reduce waste of materials. A further consideration was to prevent the acceleration of deterioration in artefacts remaining in contact with poor quality materials for an indefinite period.

Objects or containers holding objects were sealed in clear polyethylene bags of various modular sizes, secured with an elastic band. The closing system worked well as it avoided the use of adhesive tape except with plastic sheeting. As objects returned from treatment to their designated space, they remained bagged to prevent re-contamination, but were reasonably visible and accessible for curatorial work, which continued throughout the project. Bags could be opened and easily re-sealed to protect treated materials from exposure to insects from untreated areas in storage. Hundreds of bags had to be opened and resealed quickly during the fumigation process. This involved opening the bags as the artefacts were placed in the fumigation chamber to allow exposure to the phosphine gas.

When the fumigation process was complete, the chamber was opened and the bags were immediately re-sealed. Boxes or trays were used to organize materials more efficiently. The main means of transit involved multi-level metal and wood trolleys and flatbed dollies. Large foam-core boards were used on the trolleys to provide a surface to accommodate large objects, such as wooden figures and architectural artefacts.

Treatment

The general policy was to treat all artefacts in order to avoid the risk of overlooking undetected eggs or concealed adults. The two main insect pests, moths and dermestids, feed primarily on organic materials composed of keratin and chitin supplemented by sugars and proteins found in soiled substances. The ethnology storage offered a huge range of food sources, including wool, silk, feather, animal hair as well as other proteinaceous materials such as baleen and quills. Many of the artefacts were impregnated with oils or were otherwise soiled. Most wooden artefacts are vulnerable to the active woodboring beetle, and infestations were discovered during the systematic examination. Infestation was so prevalent that a selective process was not a worthwhile option and very few materials could be discounted. It is known that some stone, metal and ceramic objects can contain traces of organic substances, which are potential lures for insects, and crevices and dark recesses are attractive breeding sites. Webbing clothes moths were, in fact, discovered under stone tools and together with dermestids in ceramic vessels. Therefore, all artefact materials, inorganic and organic, received the same scrutiny.

The freezing process effectively kills insects by causing cellular damage as a result of a single or combined mechanism, which includes dehydration, intracellular ice formation, leading to loss of bound water, and reduced enzyme activity. The temperature at which this occurs is dependent on the length of exposure and the ability of different species to tolerate low temperature. A temperature of −18°C is reported as lethal to the webbing clothes moth (Strang, 1992). Collected data also shows that various dermestid beetles are killed at −18°C. A rapid rate of cooling has been recommended as a means to thwart the development of freeze-resistance and the immediate repetition of the freeze–thaw cycle is another means of eliminating the ability of insects to acclimatize or supercool as if for hibernation (Florian, 1986).

Bagged artefacts were placed in two block freezers in a stacked arrangement assembled with platforms and styrofoam blocks. These were spaced to allow air circulation, in order to achieve the most rapid rate of cooling to −18°C. The variable insulating qualities of the artefact materials and the freezer density are important factors which affect the rate of cooling. After 24 hours, the objects were removed and left to thaw for a further 24 hours. This was followed by a repeat of the freeze–thaw cycle. Artefacts with slight risk factors were wrapped in layers of cotton muslin inside the plastic bag to moderate changes.

Cycled freezing was the preferred option to fumigation with phosphine (Phostoxin™), because the application of a chemical treatment is a potential health risk to handlers, and has the potential for interaction with artefact materials and previously applied chemicals, with unknown consequences. It may also adulterate any material analyses that could be done in the future.

However, the fumigation treatment was used to treat oversize objects that exceeded the dimensions of the freezer, and objects judged to be vulnerable to damage during the freeze–thaw process. These included certain types of surface coatings, deteriorated ivory, teeth, wax components, materials under tension (such as stretched skin elements) and objects with developing or recent cracks. A further concern was that the minimum temperature capability of the freezers in conjunction with the exposure time schedule was not certain to kill all stages of woodborer beetles, particularly those present in the core.

The fumigation treatment of artefacts entailed a two-week scheduling, which involved a multi-stage process. First, the artefacts were transported and then placed, with bags opened, in a large fumigation chamber situated in the basement. Next, a licensed pest control firm carried out the application of the pesticide, followed by venting and testing the chamber. The fumigant was applied at 20°C at 200 ppm for 72 hours. When the fumigation process was finished an extra margin of time was allowed for off-gassing whenever possible. The team then re-sealed the bags containing the artefacts in the chamber and moved them back into storage. Although phosphine has no residual insecticidal properties, one disadvantage of Phostoxin™ is its reaction with copper metals and alloys, which can result in corrosion and discoloration. Therefore, all copper-containing components of an object were protected from exposure to phosphine during fumigation by the application of a coating of 3% benzotriazole in ethanol (Fenn, 1988). Green and blue paint layers, stains and some dyes were tested during the examination process to determine the presence of copper.

Alternate strategies

Vulnerable inorganic objects, that could not be treated safely or were considered too low risk to justify moving, were examined thoroughly and if pest-free were sealed in plastic and marked 'bagged only'.

The treatment developed for medicine bundles from the North American Plains was influenced by cultural beliefs. One of the beliefs upheld is that the contents of many medicine bundles should not be viewed except by the person who has the ceremonial right to use that bundle. Therefore these were not opened. The inability to examine the condition of the contents meant that the treatment options discussed above could not be carried out on technical and ethical grounds. Nor was the application of chemicals such as metallic phosphide or permanent enclosure thought to be appropriate.

Catalogued descriptions showed that the wrapped bundles included a wide range of proteinaceous materials, which would be prone to damage from insect pests, especially those which favour dark habitats and undisturbed conditions. Since the possibility of visual monitoring for pests was not an option, a solution was developed which involved the permanent relocation of the medicine bundles to a cold storage facility at a constant temperature of 5°C. Low temperatures are known to effectively restrict feeding and insect development (Strang, 1992).

The quarantine time recommended after sealing and treatment of the entire collection was a minimum of three months after the completion of the general cleaning of the storage room. The plastic bagging on some artefacts was retained as protective sleeves.

OBSERVATIONS

The insect species found in ROM ethnology objects and storage room include:

- webbing clothes moth (*Tineola bisselliella*)
- case-making clothes moth (*Tinea pellionella*)
- black carpet beetle (*Attagenus unicolor*)
- odd beetle (*Thylodrias contractus*)
- red flour beetle (*Tribolium* species)
- American cockroach (*Periplaneta americana*)
- psocid or booklouse
- woodboring beetle, species undetermined

Live dermestids were often, but not consistently, found in combination with moth activity

Tracking the infestation

A spatial analysis of locations and sightings of insect pests in storage show two distinct features: patterns of localized insect activity and an overall distribution of sites.

Predictably, many objects stored close to the major live infestation sites were affected. A webbing clothes moth population tends to concentrate around a relatively small area. Undisturbed populations of webbing clothes moths usually breed new generations within a 2–3 m radius of the infestation core, emerging only when the food source is consumed (Norton, 1996). Moths are not strong flyers and neither larvae nor female adults laden with eggs move far from their initial food source. Unfortunately, materials equally attractive to the moths were typically grouped in close proximity in storage. In conducive conditions, i.e. readily available food source and temperatures of 22°C or higher, webbing clothes moths and case-making moths can complete their life cycle in 65–90 days. On average, the female lays approximately 40–50 eggs (Pence, 1966). Under favourable conditions, carpet beetles can produce one to four generations a year, depending on species (Edwards *et al.*, 1980).

The number of widespread infestation cores was more unsettling. The main infestation sites were situated at opposite ends of the room with other infestations distributed throughout. This may have resulted from movement of infested objects within the room, the reuse of contaminated packing materials, or from introduction from new acquisitions or artefacts returned from photography, display or loan. An investigation of past relative humidity levels in storage was prompted by the findings of large numbers of live juvenile psocids (booklice) in frass collected from a large reed boat. Psocids feed on fungal spores and mildew associated with starch-containing materials, such as plant fibres (Morris, 1997). The new live generation found together with thousands of dead adults, but no in-between stages, suggests that humidity levels had increased within the previous year sufficiently to stimulate mould growth. The efficiency of the treatments used was evaluated by two routines: visual inspection of artefacts immediately after treatment and through periodic inspections of treated areas and previously infested materials. Inspections found no live insects inside bags of treated artefacts or in storage room traps as the project neared completion.

The effectiveness of the fumigation treatment for objects with active woodboring beetles can only be assessed over a period of one to five years, depending on species, and so objects should remain sealed in plastic sheeting and regularly monitored for signs of activity. In general, treatments did not appear to have any damaging effects on materials. However, the following change was noted: widening of cracks in a thin-walled Mexican gourd, after one freeze cycle.

The ROM project required a team of two full-time staff for 12.5 months, together with assistance from students and volunteers, equivalent to one additional full-time worker for six months. Over 43,000 objects were treated in three different areas.

Post-project recommendations

Large-scale infestation in several different locations within the room indicates the need for more effective monitoring. The discovery of an infestation requires immediate action to prevent it getting out of control.

The main preventive strategies which were recommended were the following:

- *Introduction of a quarantine system.* This involves isolating and inspecting/treating all materials including packing materials coming from uncontrolled environments, e.g. those returning from loans, mounts, new acquisitions or those under consideration, exhibition materials and objects returning from photography. This is perhaps the most critical practice for preventing infestation.
- *Monitoring of collection and environment.* This involves regular inspections and use of traps, both unbaited and baited with pheromones.
- *Maintenance* of good seals around doors and installation of insect screens on ventilation ducts.
- *Upgrade of storage furniture.* Many vulnerable objects such as textiles would benefit from the protection of metal insect-proof cabinets.
- *A regular cleaning regime to remove dust and debris.* This disturbs insect habitats and removes eggs and food sources.
- *An early response* to contain activity.

In general, the same strategies apply to protecting the museum at large. The implementation of an integrated pest management programme is a valuable safeguard to reduce this type of non-recoverable damage to artefacts. The general emphasis is on prevention rather than recovery measures (Linnie, 1996), focusing on two key objectives: blocking the entry of pests and regular monitoring. In addition, the use of low temperatures or low oxygen environments limits the use of chemicals.

CONCLUSIONS

This was a pilot project for the museum. The process was labour-intensive, due in part to efforts to avoid compromising artefact safety for the wide range of materials involved. A positive result of the project is the improved storage conditions for many artefacts due to the inclusion of unobtrusive protective supports made from materials with long-term stability. The examination of all the objects in storage was an opportunity to alleviate overcrowding, and to remove decomposing packing materials. Also, the systematic examination of the entire collection revealed problems requiring future conservation attention.

The removal of insect traces and debris from artefacts would have been a further benefit, but unfortunately was not carried out due to time restraints. Apart from the aesthetic considerations, it clarifies the condition of the artefact with respect to future insect activity, as residual debris may be misleading.

REFERENCES

Edwards S, Bell B, King M E, *Pest Control in Museums: A Status Report,* 1980, The Association of Systematics Collections.

Fenn J, 'Fumigation with hydrogen phosphide "Phostoxin" at the Royal Ontario Museum', in *Proceedings of the 14th Annual IIC-CG Conference,* 1988, Ottawa, Canada, 115–123.

Florian M L, 'The freezing process – effects on insects and artifact materials', in *Leather Conservation News,* 1986, **3**(1), 1–13, 17.

Linnie M, 'Integrated Pest Management: a proposed strategy for natural history museums', in *Museum Management and Curatorship,* 1996, **15**(2), 133–143.

Morris D, 'Personal communication', 1997, Entomology, Centre for Biodiversity and Conservation Biology, Royal Ontario Museum, Ontario, Canada.

Norton R, 'A case history of managing outbreaks of webbing clothes moth (*Tineola bisselliella*)', in *Pre-prints, 11th Triennial Meeting, ICOM Committee for Conservation, Edinburgh, Scotland,* J. Bridgland (Editor), 1996, James and James Science Publishers, London, **1**, 61–67.

Pence R, *Analyzing Fur Damage with a Microscope,* California Agriculture Experimental Station, 1966, Ext. Circ. 514, University of California, California, USA.

Strang T J K, 'A review of published temperature for the control of pest insects in museums', in *Collection Forum,* 1992, **8**(2), 41–67.

MATERIALS AND EQUIPMENT

Ethafoam® and polyethylene foam
    The Dow Chemical Company
    Midland
    Michigan 48674
    USA

Polyethylene bags
    VG Enterprises
    Woodbridge
    Ontario
    Canada
    Tel: +1 905 851 0794

Chiswick Packaging Products for Industry, Businesses and Retail
    PO Box 4444
    Georgetown
    Ontario L7G 4X7
    Canada
    Tel: +1 800 225 8708
    Fax: +1 800 526 0066

Coroplast® (lightweight corrugated polypropylene/ polyethylene copolymer sheet), museum mounting board and unbuffered Abaca fibre tissue
    Woolfitt's Art Enterprises, Inc.
    1153 Queen St. West
    Toronto
    Ontario M6J 1J4
    Canada
    Tel: +1 416 536 4322
    Fax: +1 416 536 4322

Cotton twill tape
    Dressmakers' Supply
    1110 Yonge St.
    Toronto
    Ontario
    Canada
    Tel: +1 416 922 6000

ACKNOWLEDGEMENTS

The author gratefully acknowledges the support and technical help contributed by Julia Fenn, Ethnographic conservator, ROM. The support of Dr Mima Kapches and the members of the Anthropology Department and all who participated in the project. Dael Morris of Entomology, ROM, is thanked for her prompt identification of insects through out the project.

BIOGRAPHY

Following studies at the University of Toronto, Elizabeth gained a MAC in the Master of Art Conservation programme at Queen's University at Kingston, Canada in 1995, specializing in conservation of objects. After completing an internship at the Metropolitan Museum, New York, she has carried out conservation projects as Cultural History conservator at the Museum of Natural History and Archaeology, the Norwegian University of Science and Technology in Trondheim, Norway, the Royal Ontario Museum in Toronto, Canada, and at the Chicago Historical Society as a Getty Intern.

Pest control in the Greek region

Elisa Polychroniadou and Athanasios Mitseas

13 Omirou Street, N Psychiko, 15451 Athens, Greece
e-mail: epolychroniadou@hotmail.com

The poster presented various types of insect pests found in the Greek region, which attack organic materials such as wood, paper and textiles. The various methods for control, including the preventive approach and methods of eradication, are described. Insect pests collected from specific areas and under specific conditions from the Dodecanese and Cycladic islands and mainland Greece were studied.

Photographs of insect pests were presented including specimens from the following locations:

- Church of Panagia Ammou in Sifnos
- Ioannis Chrysostomos and Agios Spyridonas on the island of Sifnos
- Museum of Decorative Arts and the Archaeological Institute
- Castle of Magistrate on the island of Rhodes
- Chatziagapitou room, and the Chaviara archives on the island of Symi
- Museum of Greek Traditional Instruments
- Fivos Anoyiannakis collection

In addition, various examples from private collections and residencies were also presented.

New technologies and ethical issues were raised, highlighting concern for the use of insecticides, fumigants and their effects on humans and the environment. New types of insects, their multiplication and their resistance to pesticides due to the changing environmental conditions in Greece, i.e. rise of temperature and humidity, were also highlighted.

ANALYSIS OF THE PESTICIDE RESIDUES PRESENT ON HERBARIUM SHEETS WITHIN THE NATIONAL MUSEUMS AND GALLERIES OF WALES

Victoria Purewal

*Department of Biodiversity and Systematic Biology, National Museums and Galleries of Wales,
Cathays Park, Cardiff CF10 3NP, Wales
Tel: +44 29 205 73224   Fax: +44 29 202 39829   e-mail: marc@arsenalfc.net*

The organic components in natural history collections are particularly prone to pest and mould attack. To prevent the destruction that biological deterioration will impart, chemicals toxic to these organisms have been applied, such as inorganic arsenic trioxide, mercuric chloride, barium fluorosilicate and organic pesticides including lindane, DDT and naphthalene.

A major cause for concern was the stability of the inorganic residues over long periods of time. Analysis of specimens in the herbarium at the National Museums and Galleries of Wales (NMW herbarium) disclosed that the majority of the collection still had high concentrations of mercury, arsenic and barium present on the sheets. No inorganic application had been made in at least 70 years. The concentration of mercury in some cases was as high as 1000 ppm, enough to pose a serious threat to the health and safety of staff who handled and worked with the collections. Curators had been completely unaware of its presence before this research had been conducted.

Several methods of analysis are available for the identification of residues but the most effective for this study were Inductively Coupled Plasma Mass Spectrometry (ICP-MS), Flow Injection Mercury System Atomic Emission Spectrometry (FIMS-AES) and Atomic Absorption Spectrophotometry (AAS).

Mercury and arsenic are highly toxic metals that can accumulate in the body, producing serious acute and chronic symptoms. Biological monitoring was initiated, involving the analysis of blood and urine to determine whether the staff members handling the collections had been contaminated. Two members of staff were found to have higher concentrations of both arsenic and mercury, but these levels were still within the Health and Safety Executive (HSE) guidelines. Monitoring the air quality of the herbarium showed that the concentration of mercuric chloride vapour was well within the HSE guidelines due to the adequate ventilation within the herbarium. After just one year of monitoring and the implementation of safe handling techniques, staff members' mercury and arsenic concentrations had all returned to normal.

RECOMMENDATIONS

To ensure personal safety when working in an environment housing historic collections, the following recommendations are given:

- monitor the air quality
- undertake health surveillance checks every 6–12 months
- wear nitrile gloves when handling specimens directly
- wear dust masks to prevent inhalation of particulate material
- ensure the working environment is well ventilated
- wash hands after handling any material

VELOXY®: A NOVEL SYSTEM FOR NITROGEN DISINFESTATION TREATMENTS

Simon Conyers

Pest Management Group, Central Science Laboratory, Sand Hutton, York YO41 1LZ, United Kingdom
Tel: +44 1904 462048   Fax: +44 1904 462252   e-mail: s.conyers@csl.gov.uk

Veloxy® is a method of on-site nitrogen generation, which has been devised for insect control on museum artefacts. It removes the need for the use of bulky nitrogen cylinders and treatment chambers or pre-formed bubbles. With the development of Veloxy®, particular care has been taken to produce an easy to operate robust system, which is readily transportable and could make use of the domestic power supply. It has been developed through a project entitled 'Save Art', funded by the European Union. There were six different partners in the consortium:

- two small industrial enterprises from Italy responsible for developing and building Veloxy® and the monitoring devices
- one UK government laboratory doing efficacy tests on the nitrogen atmospheres against a range of museum pest species
- three museums from Italy, Spain, and Sweden responsible for testing the system on a range of artefacts

The Veloxy® system consists of:

- a portable 1.1 kW compressor with dimensions of 0.70 m length, 0.26 m width and 0.83 m height
- the Veloxy® (Resource Group Integrator s.r.l., via Nazario Sauro 8, 16145 Genova, Italy; e-mail: rgi@mbox.ulisse.it) portable permeation system of dimensions of 0.37 m x 0.40 m x 0.94 m
- relative humidity generation system
- plastic sheeting, heatsealer, tubing and inlet and outlet ball valves to create enclosure
- oxygen meter to monitor conditions within the enclosure

The enclosure is evacuated and then the output from the Veloxy® is attached. For effective insect control, oxygen within the treatment enclosure is reduced to 0.3% or lower. This process can be accelerated by recycling the output from the enclosure back into the compressor. This is a proven system which has been designed for use by staff involved with collections.

THERMO LIGNUM® APPLICATION OF WARMAIR AND NOXIA: COMPLETE NON-CHEMICAL PEST ERADICATION

Karen Roux and Paul Leary

Thermo Lignum UK Limited, 19 The Grand Union Centre, West Row,
London W10 5AS, United Kingdom
Tel: +44 20 8964 3964   Fax: +44 20 8964 2969   e-mail: thermolignumuk@compuserve
Website: http://www.thermolignum.com

Thermo Lignum® has been applying the WARMAIR method to an increasing array of organic materials which can come under attack from insects. WARMAIR consists of the computer-controlled application of elevated temperatures (up to 55°C). It achieves a complete kill of the infestation life cycle stages, whilst maintaining constant relative humidity (RH) throughout the process. Maintaining the moisture equilibrium between the material components of the object and the ambient air during treatment ensures that there can be no physical change in the object (Ertelt, 1993). The elevated temperature guarantees the complete eradication of all insect pests (Pinniger, 1996).

In order to broaden the application of this process to objects, which due to their age, conservation history, fragility, or for practical reasons have previously presented a problem in terms of the treatment method, Thermo Lignum® has successfully experimented with a combination of the WARMAIR method at lower temperatures (ambient to 36°C) with anoxic atmospheres. This development work resulted in the NOXIA method.

Whilst a lower treatment temperature, by itself, is not sufficient to achieve a complete eradication, it aids and accelerates the known effect of a virtually oxygen-free atmosphere. NOXIA enables the conservator to select the temperature which is considered safe for the object whilst also reducing the treatment time needed for conventional anoxic treatments.

The poster illustrated the two treatment techniques using photographs of objects in the WARMAIR chamber and others being prepared for a NOXIA treatment. Both processes were further described using typical computer printouts, which outlined temperature and RH measurements over the course of the treatment.

REFERENCES

Ertelt P, *Studies on Controlled Thermal Treatment of Pest Infested Wood,* Thesis (Diplomarbeit), 1993, Rosenheim Technical College, Department of Wood Technology.

Pinniger D, 'Insect control with the Thermo Lignum® treatment', in *Conservation News,* 1994, **59**, 27–28.

BISCUIT BEETLE: SAFE HAVENS AND DANGEROUS DISPLAYS

Lynn Morrison
CONSERVATION OFFICER/CARE OF COLLECTIONS
Saffron Walden Museum, Museum Street, Saffron Walden, Essex CB10 1JL, United Kingdom
Tel/Fax: +44 1799 510333   e-mail: museum@uttlesford.gov.uk

Saffron Walden Museum opened its doors in 1835, and has a long history of pest treatment, environmental control and latterly, insect monitoring. In October 1998, an outbreak of insect activity was noted in the museum's main gallery, and a new pest was identified. Biscuit beetles *Stegobium paniceum* were found in quantities on window sills. These adult beetles from the *Anobiidae* family, are 2–5 mm long, very active, and fly towards the light (Pinniger, 1994). They attack a wide range of dried plant material including biscuits and other cereal products, tobacco, seeds, and even chocolate (Mound, 1989). There was concern that they would attack our 19th century herbarium.

After discussions with the Council's Environmental Services and an external consultant, an intensive monitoring strategy was set up and fumigation before Spring was considered. Two building reconstructions, which formed part of the displays relating to archaeology, contained large amounts of thatch in their roofs. The consultant strongly advised getting rid of this as we suspected the straw could harbour the insects and provide a food source for them. In March, a biscuit beetle fell out of the thatch onto the Chairman's head whilst he was addressing the Saffron Walden Museum Society, so removing the thatch became ever more pressing.

Thatch removal was arranged in house, and was carried out by Council workmen on 26 March 1999. The display cases were taped up and the gallery closed, but nonetheless the amount of fine, dark dust created was surprising. The next day, staff vacuumed the gallery and noticed a pronounced mouldy smell, as well as dust penetrating all surfaces.

The workmen's dust masks and overalls proved little defence against the dust, and they all quickly suffered allergic reactions involving respiratory, skin and eye irritation, needing up to six weeks off work. Due to this, a further professional clean up was required, by the sort of company which removes asbestos and other hazards.

The Health and Safety Executive, called in by the Council, appointed a Consultant Occupational Physician from Cambridge to examine the workmen and the dust. Samples were found to contain an elevated concentration of endotoxin, but no other chemical contaminants. The fungal and bacterial culture studies did not reveal any abnormal organisms and he concluded that the men suffered an unusual irritative episode linked to the high dust concentrations and their endotoxin content. Endotoxins are part of the outer cell wall of bacteria, found universally in air, water and soil, which flourish on mould and dry organic matter. Endotoxins cause a range of reactions including fever; their effects in industry and agriculture have long been known, with cases in the heritage sector now being examined and documented.

Investigations into the thatch straw showed that most of the seed heads had been removed in the threshing process. It was not treated prior to installation in 1988, apart from with a fire retardant. The thatch was too high to clean and thus had allowed dust to settle on it over the last ten years.

It is clear with hindsight that a large volume of organic material such as two straw roofs within an interior space will attract dust, bacteria, mould spores and become a sanctuary for passing insects, with unpredictable results.

Pest monitoring continues, as well as regular cleaning and occasional spraying of stores and galleries. The thatch of the roofs has not been replaced, and they have been left with their wooden sub-frames showing. What of the Biscuit beetle? It was never seen again.

REFERENCES

Mound L (Editor), *Common Insect Pests of Stored Food Products – A Guide to their Identification,* 1989, Natural History Museum, London.

Pinniger D, *Insect Pests in Museums,* 1994, Archetype Publications, London.

THE THERMO LIGNUM® METHOD OF PEST ERADICATION, WITH SPECIFIC REFERENCE TO ENTOMOLOGICAL COLLECTIONS IN THE NATURAL HISTORY MUSEUM

Phil Ackery, Adrian Doyle and David Pinniger

Entomology Department, The Natural History Museum, Cromwell Road, London SW7 5BD, United Kingdom
Tel: +44 20 7942 5612   Fax: +44 20 7942 5229   e-mail: pra@nhm.ac.uk

A recent Natural History Museum (BMNH) internal report (Harbord, 1999), concluded that:

> *Thermo Lignum® treatment could be incorporated into the pest-avoidance strategy [of the BMNH] as a one-off treatment for large zoological specimens, return of large-scale exhibition loans, new acquisitions and as part of the Disaster Plan for the treatment of water damage and in the event of major infestations.*

To this, we would add the possibility of incorporating such treatment within a strategy for moving extensive collections from one store to another, as envisaged for the planned new BMNH Entomology and Botany block, the Darwin Centre Phase II.

We identified four major criteria with respect to large-scale treatment of our entomological collections. Any method adopted would require a short turnover time and would need to be demonstrably non-detrimental to the collection containers and specimens held therein. It would also need to be logistically feasible (remembering that we are talking about 120,000 insect drawers!) and not require lids to be removed from the collection drawers for the duration of the treatment. Conventional high-temperature and low-temperature treatments, modified atmospheres and available options for chemical fumigation, all fail on one or more of these criteria.

Our attention turned to the Thermo Lignum® (Thermo Lignum® UK Ltd, UK) method which employs a high temperature regime (53°C for 24 hours), whilst uniquely maintaining relative humidity (RH) within set narrow parameters. Therefore, we could be confident that there would be no damage to the collection containers through drying-out. As to our other desired criteria, the treatment period is short compared to the three weeks required for modified atmospheres, and there is no need for the logistically difficult 'bagging-up' associated with low temperature regimes. The outstanding problem remained the fluctuations in humidity within the enclosed environment of the collection drawers. Our expectation was that the fabric of the drawer would act as a buffer to the conditions within, so that as the temperature increased, the water-bearing capacity of the air held within the drawer would increase, resulting in an unacceptable decrease in RH.

Our initial test, carried out on a single entomological drawer, involved monitoring conditions both internally and externally by means of two Tinytag Ultra7 monitors. The test demonstrated quite clearly that changes to the internal environment of the drawer closely mirrored the controlled changes within the treatment chamber itself, contrary to our expectation. Funding is being sought to support a more extensive trial utilizing a range of differently designed collection drawers. In addition, we intend investigating the difference in conditions when the drawers are stacked, or held in racks during treatment. Our expectation is that the results will be of particular relevance to our major task of moving and treating the collections, and in developing future control strategies.

REFERENCES

Hardbord R I, *Thermal Treatment Test Report,* unpublished internal document, 1999, Natural History Museum, London.

MATERIALS AND EQUIPMENT

Thermo Lignum® UK Ltd
19 Grand Union Centre
West Row
London W10 5AS
United Kingdom

Tinytag Ultra®
Meaco
Unit 8
Smithbrook Kilns
Cranleigh
Surrey GU6 8JJ
United Kingdom

THE JOURNEY OF THE LOST BISCUIT BEETLES

Suzette Hayes and Andrew Calver
CONSERVATION DEPARTMENT

Museum of London, London Wall, London EC2Y 5HN, United Kingdom

Tel: +44 20 7814 5641   Fax: +44 20 7600 1058   e-mail: shayes@museumoflondon.org.uk

The Museum of London has been running an Integrated Pest Management (IPM) programme for over five years, both at the main museum site at London Wall, and at the external object store at Eagle Wharf Road (EWR) (Pinniger *et al.*, 1998). This includes monitoring for pests using sticky blunder traps (Child and Pinniger, 1994), some with pheromone lures where applicable, regulating temperature and relative humidity to discourage pest attack, carrying out regular housekeeping, having a quarantine area and treating pest infestations promptly. The sticky traps are checked monthly and the data is logged. During these inspections occasional pests were caught, such as booklice *Liposcelis,* moths or woolly bear *Anthrenus* species casings.

However, in April 2000 a dozen biscuit beetles *Stegobium paniceum* were found on one trap in one of the storage bays at EWR, near some post-medieval wooden water pipes. These beetles are a member of the furniture beetle *Anobium* family and have a starch-containing food diet (Pinniger, 1990; Mourier *et al.*, 1977). The surrounding area was thoroughly examined and no obvious source of infestation was found. It was thought that the beetles might be attracted to the indicator light in an electrical junction box in the corner of the bay, therefore extra sticky traps were placed, including one adjacent to the light. The results confirmed the above. To test the light attraction theory further, a light trap (flea trap No. 1) was placed on the floor 3 m away from the junction box and the bay was kept in darkness. After 14 days the sticky disc in the flea trap was examined and 50 adult biscuit beetles were found, plus more on the ground around the trap. Three more flea traps were placed throughout the bay to pinpoint the source, but no beetles were found on these traps, while the numbers increased on trap No. 1. Museum staff carried out an extensive search of the bay and identified a number of pieces of bakery equipment, which contained flour residues and evidence of old biscuit beetle infestation. These objects were bagged in polythene, in situ, to contain any infestation and to help localize the source. Since bagging the objects, only three beetles were found inside the bags but there was still a steady flow of beetles on trap No. 1, indicating that these objects were not the source of the current infestation.

Having exhausted our IPM procedure, we called in David Pinniger, an Entomologist and Pest Control Consultant, to advise us on the outbreak. He reassured staff that they had carried out a logical programme of trapping and investigation and could not have done more to check and isolate suspect material. However, it was noted that there were a few beetles on top of the bagged objects, therefore the nearest hanging strip lights were checked. The two lights hanging either side of flea trap No. 1 had about 10 and 20 biscuit beetles, respectively, on the top of the fittings, all grouped near the hanging chain with the electrical wire feed, but none inside the light fittings. The other light fittings in the bay were checked but these were either clear or had one or two beetles on top. For these two light fittings, the power cable supply emerges from the ceiling suggesting a hidden electrical conduit route, as opposed to the other fittings where the cable is run externally. David Pinniger recommended that the wire feeds for these two lights should be opened to see if there was a ceiling cavity, which might contain a food source, such as old rodent bait or a mummified rodent. He concluded in saying that this was one of the strangest and most puzzling infestations he had encountered.

The results from the trapping have clearly demonstrated that biscuit beetles are attracted to the tungsten light source of the flea trap and the panel indicator light. UV readings from these sources were low, at 20 µW/lumen and 3 µW/lumen, respectively. This may suggest that biscuit beetles are attracted to visible light rather than UV, but this has yet to be confirmed. The environment in the bay is suitable for biscuit beetles to breed (19°C and 50% RH) suggesting that the source is within the bay. The next stage is to examine the ceiling cavity to hopefully reveal the long lost food source.

REFERENCES

Child R E, Pinniger D B, 'Insect trapping in museums and historic houses', in *IIC 15th International Congress, Preventive Conservation, Practice, Theory and Research,* 1994, Ottawa, Canada, 129–131.

Mourier H, Winding O, Sunesen E, *Collins Guide to Wildlife in House and Home,* 1977, Collins, London.

Pinniger D, *Insect Pests in Museums,* 1990, Archetype Publications, London.

Pinniger D, Blyth V, Kingsley H, 'Insect trapping: the key to pest management', in *3rd Nordic Symposium on Insect Pest Control in Museums,* 1998, 96–107.

MATERIALS AND EQUIPMENT

The Ultimate Flea Trap and Sticky Traps
    Killgerm Chemicals Ltd
    Denholme Drive
    Ossett
    West Yorkshire WF5 9NB
    United Kingdom
    Tel: +44 1924 277631
    Fax: +44 1924 264757

ACKNOWLEDGEMENT

The authors wish to thank Mike Ashington for carrying out the IPM at EWR and for making them aware of the situation quickly, as well as his assistance with the above work.